MOBILE AND PERSONAL COMMUNICATION SERVICES AND SYSTEMS

Books in the IEEE Press Series on Digital & Mobile Communication

DIGITAL TRANSMISSION ENGINEERING
John B. Anderson
1999   Hardcover   400 pp   IEEE Order No. PC5714   ISBN 0-7803-3457-4

FUNDAMENTALS OF CONVOLUTIONAL CODING
Rolf Johannesson and Kamil Sh. Zigangirov
1999   Hardcover   448 pp   IEEE Order No. PC5739   ISBN 0-7803-3483-3

WIRELESS COMMUNICATIONS: Principles and Practices
A Prentice-Hall book published in cooperation with IEEE Press
Theodore S. Rappaport
1996   Hardcover   656 pp   IEEE Order No. PC5641   ISBN 0-7803-1167-1

FUTURE TALK: The Changing Wireless Game
Ron Schneiderman
1997   Hardcover   272 pp   IEEE Order No. PC5679   ISBN 0-7803-3407-8

MOBILE AND PERSONAL COMMUNICATION SERVICES AND SYSTEMS

Raj Pandya

IEEE Communications Society, *Sponsor*

IEEE PRESS

John B. Anderson, *Series Editor*

The Institute of Electrical and Electronics Engineers, Inc., New York

This book and other books may be purchased at a discount
from the publisher when ordered in bulk quantities. Contact:

IEEE Press Marketing
Attn: Special Sales
Piscataway, NJ 08855-1331
Fax: (732) 981-9334

For more information about IEEE Press products, visit the
IEEE Home Page: http://www.ieee.org/press

*TK5103
,485
.P 36
1999

Copy 2*

Printed in the United States of America

10   9   8   7   6   5   4   3   2   1

ISBN 0-7803-4708-0

IEEE Order Number PC5395

Library of Congress Cataloging-in-Publication Data

Pandya, Raj, 1932–
    Mobile and personal communication services and systems / Raj Pandya.
        p.   cm. — (IEEE Press series on digital & mobile communication)
    Includes bibliographical references and index.
    ISBN 0-7803-4708-0
    1. Personal communication service systems.   2. Mobile communication
systems.   I. Title.   II. Series.   III. IEEE series on mobile & digital
communication.
    TK5103.485.P36   1999
    621.38455—dc21   99-31330
                                                                          CIP

To the memory of my parents,
Shreenath and Sita Pandya

CONTENTS

PREFACE

The idea and the scope for the book emerged from my own experience in attempting to acquire a good understanding of the rapidly evolving field of mobile and personal communication systems and standards, at a reasonable level of detail and breadth of coverage. Generally, one comes across either books that cover a range of technologies with very high-level descriptions suitable for a very broad, almost nontechnical audience, or books that address a specific technology or topic at a level of detail suitable only for a narrow audience of specialists. The book attempts to bridge this divide, with the text targeted at telecommunications professionals who are looking for a clear understanding of the basic technology, architecture, and applications associated with the current and future mobile communication systems, services, and standards. In terms of the depth and breadth of coverage, the book is therefore aimed to serve the following purposes and audiences as:

- a technical training text and guide for scientists, who expect to engage in planning and design of mobile and personal communication systems and networks for equipment vendors and network operators.

- a reference text for technical managers in areas of planning, engineering, and marketing of mobile and personal communication products and services.

- an introductory text for senior engineering and science students at the university and technical college level.

Thus, the primary objective of the book is to provide a systems engineering view of mobile and personal communication systems and services, and their evolution toward next-generation systems.

Topics like analog and digital cellular, cordless telephony, mobile data communications, global mobile satellite systems, next-generation wireless systems, and personal mobility systems constitute the mainstream components of today's mobile and personal communications scenario, and they form the bulk of the subject matter for this book. Two additional and important topics addressed in this book are numbering and identities and performance benchmarks for mobile and personal communications networks, which are directly associated with wide-area roaming and the user-perceived quality of service (QOS), respectively. Even though seamless wide-area roaming and improved QOS are emerging as key user requirements, a comprehensive treatment of numbers and identities and performance benchmarks in mobile network operations and system design is not readily available. To fill this gap, Appendixes A and B address these two topics, providing a useful source of relevant materials and applicable standards.

The book also attempts to provide the reader with some insight into relevant standardization activities associated with mobile and personal communications. The frequency spectrum is a key resource that needs to be shared and deployed efficiently and effectively by the wide range of wireless systems. Furthermore, to ensure end-to-end delivery of services, these wireless networks have to interwork and interoperate not only with each other, but also with public networks (PSTN, ISDN, PDN, Internet, etc.). Thus, international, regional, and national standards play a very important and critical role in the development of mobile and personal communication systems. The direction and shape of these emerging standards are to a large extent based on inputs from the broad wireless industry (operators, vendors, and regulators) and, therefore, reflect their collective and aggregated view. Some understanding of standardization activities around the world is therefore necessary to fully appreciate the emerging markets and technologies.

In a book like this—which addresses systems, services, and standards aspects of mobile and personal communications—frequent use of acronyms and abbreviations is almost inevitable and unavoidable. To partly alleviate the readers' frustration in dealing with this perennial problem, these terms are spelled out as often as is practical, and an extensive list of acronyms and abbreviations is provided as Appendix C.

Raj Pandya
Kanata, Ontario
Canada

ACKNOWLEDGMENTS

I would like to acknowledge the contributions and support provided by my family, friends, and colleagues toward completion of this book. Special appreciation is due to the members of my family: my wife Margaret, daughter Malini, son Ravi, and daughter-in-law Linda, for their continued support and encouragement during the ups and downs of this long journey.

Raj Pandya
Kanata, Ontario
Canada

ACKNOWLEDGMENTS

CHAPTER 1

INTRODUCTION

The underlying vision for the emerging mobile and personal communication services and systems is to enable communication with a person, at any time, at any place, and in any form, as illustrated in Figure 1.1. Besides providing unlimited reachability and accessibility, this vision for personal communications also underlines the increasing need for users of communications services to be able to manage their individual calls and services according to their real-time needs. For example, during certain period of the day a user may wish to divert his or her calls to a message center or to have the calls screened so that the incoming calls can be treated according to the user's instructions.

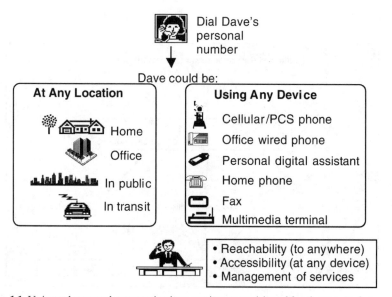

Figure 1.1 Universal personal communications services as envisioned for the twenty-first century.

1.1 ENABLING CONCEPTS FOR MOBILE AND PERSONAL COMMUNICATIONS

The concepts enabling us to provide universal personal communications include terminal mobility provided by wireless access, personal mobility based on personal numbers, and service portability through use of intelligent network (IN) capabilities. These concepts are being utilized at the national, regional, and international levels to specify and standardize a range of mobile and personal communication systems and services. As shown in Figure 1.2, the basis for terminal mobility and personal mobility is use of wireless access and personal (rather than terminal) numbers, respectively. Service portability is based on the emerging IN concepts, which facilitate real-time management of service profiles.

1.1.1 Terminal Mobility, Personal Mobility, and Service Portability

Terminal mobility systems are characterized by their ability to locate and identify a mobile terminal as it moves, and to allow the mobile terminal to access telecommunication services from any location—even while it is in motion. Terminal mobility is associated with wireless access and requires that the user carry a wireless terminal and be within the radio coverage area. Depending on the terminal design, part of the mobile terminal functions may reside on an (easily) removable and portable integrated circuit (IC) device or on a smart card.

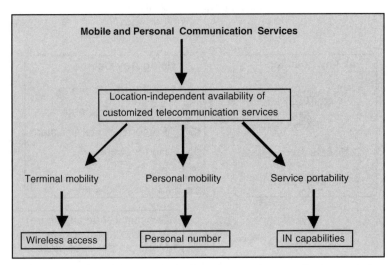

Figure 1.2 Enabling concepts for mobile and personal communication services.

The scope and applications of terminal mobility are rapidly expanding through advances in wireless access technologies and miniaturization of mobile terminals. Though the relationship between the network (line termination) and the terminal in mobile networks is a dynamic one, the relationship or association between the terminal and the user is still a *static* relationship. In other words, the communication is always between the network and the terminal, so that call delivery and billing always are based on terminal identity or mobile station number.

Personal mobility, on the other hand, relies on a *dynamic* association between the terminal and the user, so that the call delivery and billing can be based on a personal identity (personal number) assigned to a user. Personal mobility systems are therefore characterized by their ability to identify end users as they move, and to allow end users to originate and receive calls, and to access subscribed telecommunication services on any terminal, in any location. Personal mobility is not associated with any specific type of access technology (whether fixed or wireless).

The emerging implementations and integration of IN capabilities within fixed and mobile networks provide the underpinnings of a dynamic relationship between the terminal and the user. With such a dynamic association, complete personal mobility within networks and across multiple networks can be achieved. Figure 1.3 illustrates the static and dynamic relationships among the network, the terminal, and the user, and the resultant forms of mobility.

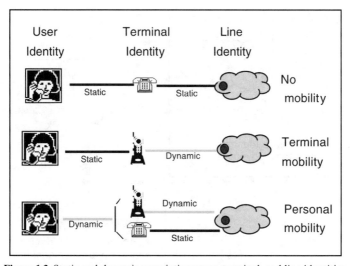

Figure 1.3 Static and dynamic associations user, terminal, and line identities.

The concepts of terminal mobility and personal mobility are further illustrated in Figure 1.4, which shows that personal mobility based on a *personal number* is a broader concept and is applicable to both wireline and wireless networks.

Service portability refers to the capability of a network to provide subscribed services at the terminal or location designated by the user. The exact services the user can invoke at the designated terminal, of course, depend on the capability of the terminal and the network serving the terminal. Service portability is accomplished through the use of IN concepts whereby the user's service profile can be maintained in a suitable database, which the user can access, interrogate, and modify to manage and control subscribed services.

1.1.2 The Intelligent Network (IN) Concept

Besides providing the capability for management and transfer of user service profiles in mobile and personal communication systems, the IN concept has a broad range of applications in fixed and wireless networks. For example, both North American and European digital mobile networks are introducing IN capabilities to provide a range of so-called IN services, which their users can access as they roam within and outside their home networks.

The term "intelligent network" describes an architectural concept that is intended to be applied to all telecommunication networks. IN aims to facilitate the introduction of new services by decoupling the functions required to

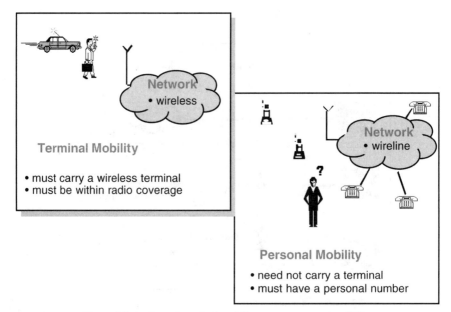

Figure 1.4 A view of terminal mobility versus personal mobility.

support call and connection control from those required to support service control, thereby allowing the two sets of functions to be placed on different physical platforms. New services can therefore be defined and implemented quickly, efficiently, and cost-effectively because major software changes to the switching systems, which were required in the pre-IN network architectures, are not necessary. IN architectural concepts are being used in a wide variety of networks including mobile and personal communication networks to capture the functions and relationships between the functions to support a set of service features and capabilities. The key attributes of IN structured networks include the following:

- extensive use of information processing techniques
- efficient use of network resources
- flexible allocation of network functions to physical entities
- communication between network functions via service-independent interfaces
- modular service creation using service-independent building blocks (SIBs)

International IN standards are being developed in phases, with increasing levels of capabilities available within the IN structured networks starting from IN Capability Set 1 (IN CS-1). Currently international standards are available for IN CS-1, IN CS-2, and IN CS-3. Some of the key services that can be supported by IN CS-1, IN CS-2, and IN CS-3 include:

- UPT (universal personal telecommunication)
- freephone (800/888 number) services
- alternate billing services
- local number portability
- virtual private networking

The basic functional architecture for IN CS-1 structured networks is shown in Figure 1.5. The various functional entities (FEs) indicated in Figure 1.5 support service execution, service creation, and service management. The role of these FEs are briefly described as follows:

Call control agent function (CCAF): provides user access capabilities and may be viewed as a terminal through which a user interacts with the network.

Call control function (CCF): provides the basic switching capabilities, which include the capability to establish, maintain, and release calls and connections.

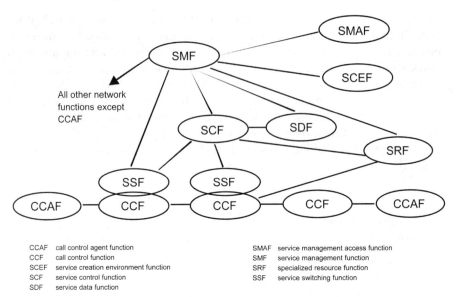

Figure 1.5 IN functional architecture.

Service switching function (SSF): cooperates with the CCF in recognizing the triggers and interacting with the service control function (SCF). CCF and SSF are considered inseparable and need to be supported in a single physical entity, but whereas CCF provides the trigger capabilities, the SSF supports the recognition of the triggers and interaction with the service control function.

Service control function (SCF): executes service logic and provides capabilities to influence call processing by requesting the SSF/CCF and other service execution FEs to perform specified actions. The key role of the SCF is to provide mechanisms for introducing new services and service features independent of switching systems.

Specialized resource function (SRF): provides a set of real-time capabilities, which may include address digit collection, announcements, text-to-voice conversion, and certain types of protocol conversion. In a physical implementation this entity is some times referred as an intelligent peripheral (IP).

Service data function (SDF): provides generic database capabilities to the SCF or another SDF.

Service creation environment function (SCEF): is responsible for developing and testing service logic programs, which are then sent to the service management function.

Service management function (SMF): deploys service logic (from SCEF) to the service execution FEs and otherwise administers these FEs by supplying user-defined parameters to customize the service and to collect billing information and service execution statistics.

Service management agent function (SMAF): acts as a terminal that provides the user interface to the SMF.

Some additional FEs introduced in IN CS-2 are the intelligent access function (IAF), the call-unrelated service function (CUSF), and the service control user agent function (SCUAF). The IAF is used to provide access to IN-structured networks from non-IN networks, and CUSF and SCUAF support call-unrelated interactions between users and service processing.

In a physical implementation, the various FEs shown in Figure 1.5 may be located in suitable physical entities (PEs) or platforms to achieve a cost-effective design. An example physical implementation of an IN-structured network is shown in Figure 1.6.

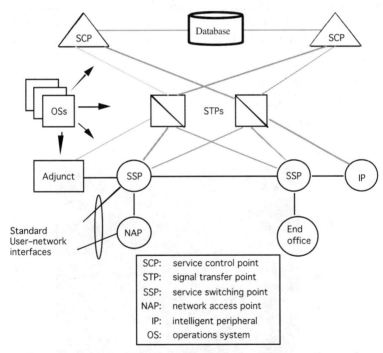

Figure 1.6 Example physical implementation of an IN-structured network.

1.2   MOBILE AND PERSONAL COMMUNICATION: PAST, PRESENT, AND FUTURE

Radio communication can trace its origin to the discovery of electromagnetic waves by Hertz in 1888 and the subsequent demonstration of transatlantic radio telegraphy by Marconi in 1901. Mobile radio systems using simplex channels (push-to-talk) were introduced in the 1920s for police and emergency services. The first public mobile radio system in the United States was introduced in 1946 and can perhaps be considered to be the beginning of the era for public mobile communication services. As illustrated in Figure 1.7, the evolution of public mobile and personal communication services may be divided into three broad periods. The development of the cellular concept in the 1970s was a defining event, which has played a significant part in the evolution of mobile communication systems and networks around the world.

The delineation of a boundary between the mobile communications in the *past* and in the *present* is not very difficult, because implementation of analog and digital cellular systems clearly represents a step change in the design and capabilities of mobile communication systems. However, a similar delineation of a boundary between the *present* and *future* mobile and personal communications systems is not so clear. Future mobile and personal communication systems will, to a large extent, represent evolution and enhancements of the present systems in many directions and on many fronts. These directions include the following:

- increased capacity and coverage
- global roaming and service delivery
- interoperability between different radio environments

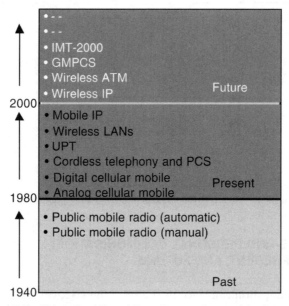

Figure 1.7 Evolution of public mobile and personal communication services.

- support of high bit rate data, the Internet, and multimedia services
- wireless wireline integration for mobile broadband services
- global coverage using satellite constellations

1.2.1   The Past

From the introduction of public mobile radio in the United States in 1946 until the first analog cellular system went into operation in Chicago in 1983, mobile radio systems were based on the *trunking* principle. In other words, the available frequency spectrum (in the 150 or 450 MHz band) was divided into a suitable number of frequency channels. A centralized, high power antenna was used to transmit signals to mobile receivers. Large mobile receivers were installed in automobiles (in the trunks), and the telephone sets also were rather large. A call originating from or terminating on a mobile terminal had to compete for one of the limited number of channels. The quality of service in terms of call blocking probabilities was very high—in the order of 20–25%. However, the users were willing to trade off the convenience of mobility against the poor quality of service in terms of call blocking and received signal quality. These systems were also severely limited in terms of capacity and coverage.

To alleviate the high blocking problem in the early systems, efforts were made to allow call originations from the mobile telephones to wait for a free channel. In the so-called automated mobile telephone system (AMTS), the mobile telephone user would key in the called number and press the *send* button. The receiver system would then start scanning for an idle channel by cycling through all the channels in the system. In some systems, the number of scan cycles was restricted, so that if an idle channel was not found within the allowed number of scans, the call would be blocked. However, incoming calls to mobile terminals (mostly originating from fixed terminals in public switched telephone networks) had no mechanism for awaiting a free channel and were blocked on all-channels-busy condition. Though the improvement in the quality of service in these systems was only marginal, they did provide some interesting performance modeling problems [04].*

1.2.2   The Present

Since the initial commercial introduction of advanced mobile phone system (AMPS) service in 1983, mobile communications has seen an explosive growth worldwide. Besides the frequency reuse capabilities provided by the cellular

---

* References of the end of each chapter are representative rather than exhaustive, and they are generally ordered according to the material covered in the Chapter. Wherever necessary and appropriate, specific references are cited in the text.

operation, advances in technologies for wireless access, digital signal processing, integrated circuits, and increased battery life have contributed to exponential growth in mobile and personal communication services. Systems are evolving to address a range of applications and markets, which include digital cellular, cordless telephony, satellite mobile, and paging and specialized mobile radio systems. Data capabilities of these systems are also coming into focus with the increasing user requirements for mobile data communications, driven by the need for e-mail and Internet access. Whereas the analog cellular mobile systems fall in the category of *first-generation* mobile systems, the digital cellular, low power wireless, and personal communication systems are now perceived as *second-generation* mobile/PCS systems.

The first digital cellular system specification was released in 1990 by the European Telecommunications (ETSI) for the global system for mobile communication (GSM) system. The GSM, DCS 1800 (1800 MHz version of GSM), and DECT (digital enhanced cordless telecommunications) systems developed by ETSI form the basis for mobile and personal communication services not only in Europe but in many other parts of the world including North America. The number of GSM subscribers worldwide exceeds 100 million and is growing rapidly.

In the United States, the implementation of digital cellular standards developed by the Telecommunications Industry Association (TIA) is progressing at a rapid rate. These standards are based on time- and code division multiple access (TDMA and CDMA) technologies. Unlike GSM, the systems are designed to operate with dual-mode terminals that can also support analog AMPS service. The intent of the emerging PCS standards in the United States is to provide a combination of terminal mobility, personal mobility, and service portability to the end users utilizing a range of wireless technologies and network capabilities. The cellular mobile and PCS standardization activity in the United States reflects the highly competitive and open-market view of mobile and personal communication services and their evolution. Rather than a single standard agreed across the entire industry, multiple standards for radio systems and network implementations have emerged, and as expected, need for marketplace and end-user acceptance is driving the ultimate implementation decisions by the operators.

A third digital cellular system called the personal digital cellular (PDC) was developed in Japan and is in full commercial operation in that country. To a large extent, the specifications for these second-generation cellular systems are being developed to meet the business and regulatory requirements in specific countries and/or regions, leading to incompatible systems that are unable to provide global mobility.

Analog cordless telephones have been in common use in residential applications, where the telephone cord is replaced by a wireless link to provide terminal mobility to the user within a limited radio coverage area. Low power digital cordless telecommunication systems like CT2 (Cordless Telephony 2), DECT, and Japan's PHS (personal Handyphone System) are intended to provide terminal mobility in residential, business, and public access applications where the users can originate and receive calls on their portable terminals as they change locations and move about at pedestrian speeds within the coverage area. It is also anticipated that the same terminal can be used in all three application environments: at the residence, at the workplace, and at public locations (airports, train and bus stations, shopping centers, etc.).

While the initial focus of the current generation of mobile and personal communication systems has been circuit-switched voice and low bit rate data services, the demand for wide-area as well as local-area wireless data services is rapidly increasing. Reflecting market needs for better mobile data services are such standards as cellular digital packet data (CDPD) for support of packet data services on analog cellular networks, high speed circuit-switched data (HSCSD), and general packet radio service (GPRS) for GSM, and IEEE 802.11 and HIPERLAN (high performance European radio LAN) for wireless LANs. The emerging industry view is that the main drivers for next-generation wireless networks will be Internet and multimedia services. Evolution toward high bit rate packet mode capabilities is therefore a key requirement for present and future mobile and personal communication systems.

With respect to personal mobility services, such features as call forwarding, call waiting, automatic credit card calling, and personal number services represent ad hoc attempts by telecommunications network operators to provide a level of personal mobility to the users. Universal personal telecommunication (UPT), the emerging standard in the International Telecommunication Union's Telecommunications Standardization Sector (ITU-T) for personal mobility, will utilize the IN and integrated services digital network (ISDN) capabilities to provide network functions for personal mobility.

1.2.3   The Future

With the rapidly increasing penetration of laptop computers, which are primarily used by mobile users to access Internet services like e-mail and World Wide Web (WWW) access, support of Internet services in a mobile environment is an emerging requirement. Mobile IP is an Internet protocol that attempts to solve the key problem of developing mechanism that allows IP nodes to change physical location without having to change IP address, thereby offering so-called "nomadicity" to Internet users.

Asynchronous transfer mode (ATM) is now generally accepted as the platform for supporting end-to-end, broadband multimedia services with guaranteed quality of service (QOS). Wireless ATM (WATM) aims to provide an integrated architecture for seamless support of end-to-end multimedia services in the wireline as well as the wireless access environment. Thus, WATM is expected to meet the needs of wireless users who are looking for a common networking solution that can meet their high speed data and multimedia service requirements with excellent reliability and service quality.

To complement the cellular and personal communication networks, whose radio coverage will be confined to populated areas of the world (less than 15% of the earth's surface), a number of global mobile satellite systems are in advanced stages of planning and implementation. These systems are generally referred as global mobile personal communications by satellites (GMPCS). GMPCS systems like Iridium, Globalstar, and ICO use constellations of low earth orbit (LEO) or medium earth orbit (MEO) satellites and operate as overlay networks for existing cellular and PCS networks. Using dual-mode terminals, they will extend the coverage of cellular and PCS networks to any and all locations on the earth's surface. On the other hand, a LEO satellite system like Teledesic aims to provide high capacity satellite links to enable delivery of high bitrate and multimedia services to every location on the earth.

International Mobile Telecommunications -2000 (IMT-2000) is the standard being developed by the ITU to set the stage for the *third generation* of mobile communication systems. The IMT-2000 standard not only will consolidate under a single standard different wireless environments (cellular mobile, cordless telephony, satellite mobile services), but will also ensure global mobility in terms of global seamless roaming and delivery of services. ETSI is also developing a third-generation mobile communication system called Universal Mobile Telecommunication System (UMTS), which will belong to the family of IMT-2000 systems.

1.3   MOBILE AND PERSONAL COMMUNICATION: SOME RELATED NETWORK ASPECTS

As the mobile and personal communication services and networks evolve toward providing seamless global roaming and improved quality of service to its users, the role of such network aspects as numbering and identities and quality of service will become increasingly important. Well-defined standards in these areas, as well as network performance for the present and future mobile and personal communication networks, will need to be addressed.

To support mobility management functions in mobile communication networks, and to provide national and international roaming, well-defined, standardized subscriber/terminal numbers and identities are required. Some form of station equipment identities is also needed to ensure that service can be denied to non-type-approved or fraudulent terminals. ITU-T plays a key role in providing these standards and is currently revising existing standards and developing new ones to support a range of terminal and personal mobility services as wireless networks evolve into the next century. An appreciation of the role of numbering and identities in mobility management, international roaming, call delivery, and billing and charging is important in understanding the operation of mobile and personal communication networks.

With the rapidly increasing penetration of wireless access in the worldwide telecommunication networks, users expect improved quality of service from mobile and personal communications systems and networks. The expectation is that future mobile communications networks will provide QOS equivalent to wireline networks. To meet this expectation, it is essential that suitable performance standards be developed and utilized in the design of wireless networks. These benchmarks are required for such areas as traffic, reliability, and transmission and generally apply not only to performance of individual network elements (like MSC, HLR, VLR) but also to the end-to-end performance of mobile and personal communication networks.

1.4   REFERENCES

[01] R. Pandya, "Emerging Mobile and Personal Communication Systems," *Communications Magazine,* Vol. 33, No. 6, June 1995.

[02] ITU-T Recommendations Q.1204 "Intelligent Networks-Distributed Functional Plane," Geneva, 1996.

[03] A. Mehrotra, *Cellular Radio: Analog and Digital Systems,* Artech House, Boston, 1994.

[04] R. Pandya and D. M. Brown, "Performance Modeling for an Automated Public Mobile Telephone System," IEEE International Communications Conference, Philadelphia, June 1982.

[05] W. C. Y. Lee, *Mobile Communication Engineering: Theory and Applications,* McGraw-Hill, New York, 1997.

[06] D. J. Goodman, *Wireless Personal Communications Systems,* Addison-Wesley, Reading, MA, 1997.

[07] M.-C. Chow, *Understanding Wireless: Digital Mobile, Cellular & PCS,* Andan Publisher, Holmdel, NJ, 1998.

2

THE CELLULAR CONCEPT AND ITS INITIAL IMPLEMENTATIONS

2.1 THE CELLULAR CONCEPT

The cellular concept was developed and introduced by the Bell Laboratories in the early 1970s. One of the most successful initial implementations of the cellular concept was the advanced mobile phone system (AMPS), which has been available in the United States since 1983. A cellular system is generally characterized as:

> a high capacity land mobile system in which available frequency spectrum is par-
> titioned into discrete channels which are assigned in groups to geographic cells
> covering a cellular Geographic Service Area (GSA). The discrete channels are
> capable of being reused in different cells within the service area.

Thus the principle of cellular systems is to divide a large geographic service area into cells with diameters from 2 to 50 km, each of which is allocated a number of radio frequency (RF) channels. Transmitters in each adjacent cell operate on different frequencies to avoid interference. Since, however, transmit power and antenna height in each cell are relatively low, cells that are sufficiently far apart can reuse the same set of frequencies without causing cochannel interference. The theoretical coverage range and capacity of a cellular system are therefore unlimited. As the demand for cellular mobile service grows, additional cells can be added, and as traffic demand grows in a given area, cells can be split to accommodate the additional traffic. Figure 2.1 illustrates an idealized view of a cellular mobile system, where cells are depicted as perfect hexagons.

A cellular system should provide the capability to hand off calls in progress, as the mobile terminal/user moves between cells. As far as possible the handoff should be transparent to the user in terms of interruption and/or call failure. Handoff between channels (in the same cell) may also be required for other reasons (e.g., load balancing, emergency call handling, meeting requirements on transmission quality).

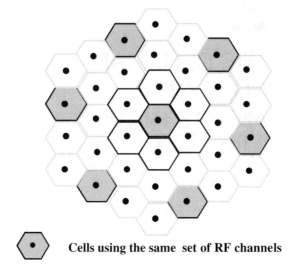

Cells using the same set of RF channels

Figure 2.1 Concept of frequency reuse in cellular mobile systems.

2.2   MULTIPLE ACCESS TECHNOLOGIES
FOR CELLULAR SYSTEMS

Generally a fixed amount of frequency spectrum is allocated to a cellular system by the national regulator (e.g., in the United States, the Federal Communications Commission). Multiple-access techniques are then deployed so that many users can share the available spectrum in an efficient manner. Multiple-access systems specify how signals from different sources can be combined efficiently for transmission over a given radio frequency band and then separated at the destination without mutual interference. The three basic multiple access methods currently in use in cellular systems are:

- frequency division multiple access (FDMA)
- time division multiple access (TDMA)
- code division multiple access (CDMA)

The underlying principle for each of these multiple-access methods is illustrated in Figure 2.2.

   In case of FDMA, users share the available spectrum in the frequency domain, and a user is allocated a part of the frequency band called the traffic channel. The user's signal power is therefore concentrated in this relatively narrow band in the frequency domain, and different users are assigned different traffic (frequency) channels on a demand basis. Interference from adjacent

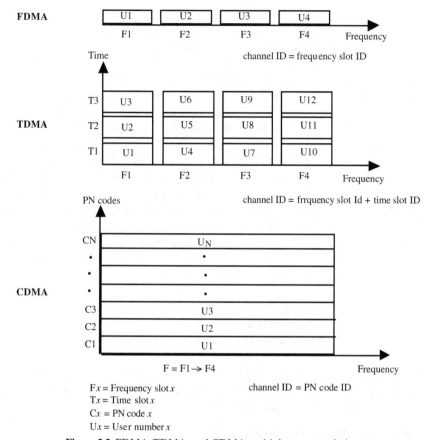

Figure 2.2 FDMA, TDMA, and CDMA multiple-access techniques.

channels is limited by the use of guard bands and bandpass filters that main-tain separation of signals associated with different users. Analog systems, which represent the initial implementations of the cellular concept, described later in this section, all use FDMA techniques.

In TDMA techniques that are utilized in many digital cellular systems, the available spectrum is partitioned into narrow frequency bands or frequency channels (as in FDMA), which in turn are divided into a number of time slots. An individual user is assigned a time slot that permits access to the frequency channel for the duration of the time slot. Thus, the traffic channel in case of TDMA consists of a time slot in a periodic train of time slots that make up a frame. In case of the North American digital cellular standard IS-136, each fre-quency channel (30 kHz) is divided into three time slots, whereas for the Eu-ropean digital cellular standard GSM, each frequency channel (200 kHz) is divided into eight time slots (full rate). In case of TDMA systems, guard bands are needed both between frequency channels and between time slots.

The CDMA systems utilizes the spread spectrum technique, whereby a spreading code (called a pseudo–random noise or PN code) is used to allow multiple users to share a block of frequency spectrum. In CDMA cellular systems (e.g., IS-95 in the United States) that use direct sequence spread spectrum techniques, the (digital) information from an individual user is modulated by means of the unique PN code (spreading sequence) assigned to each user. All the PN-code-modulated signals from different users are then transmitted over the entire CDMA frequency channel (e.g., 1.23 MHz in case of IS-95). At the receiving end, the desired signal is recovered by despreading the signal with a copy of the spreading sequence (PN code) for the individual user in the receiving correlator. All the other signals (belonging to other users), whose PN codes do not match that of the desired signal, are not despread and as a result, are perceived as noise by the correlator. Since the signals in the case of CDMA utilize the entire allocated block of spectrum, no guard bands of any kind are necessary within the allocated block.

2.3   CELLULAR SYSTEM OPERATION
AND PLANNING: GENERAL PRINCIPLES

The exact system operations and radio system planning will largely depend on the technical standards used by the mobile system. However, the following general principles are broadly applicable.

2.3.1   System Architecture

As mentioned earlier, the coverage area of a cellular system is partitioned into a number of smaller areas or cells with each cell served by a base station (BS) for radio coverage. The base stations are connected through fixed links to a mobile switching center (MSC), which is a local switching exchange with additional features to handle mobility management requirements of a cellular system. To accommodate the dynamic nature of terminal location information and subscription data, the MSC interacts with some form of database that maintains subscriber data and location information. MSCs also interconnect with the PSTN, because the majority of calls in a cellular mobile system either originate from or terminate at fixed network terminals. Figure 2.3 shows a typical cellular system architecture.

Based on the frequency spectrum made available by the licensing authority and the cellular standard in use, the cellular system is able to define a number of radio channels for use across its serving area. The available radio channels are then partitioned into groups of channels, and these groups of channels

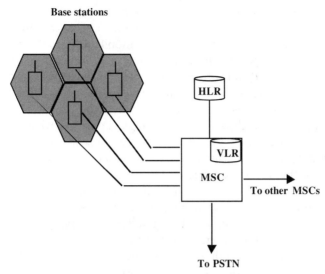

Figure 2.3 Typical cellular system architecture.

are allocated to individual cells forming the entire serving area. Individual channels or a particular group of channels can be reused in cells that are located far enough apart. A key feature of radio system planning activity consists in designing the cell sizes, assigning their locations, and allocating radio channels to individual cells. Whereas definition of channels for assignment to individual cells within a cellular system is straightforward in the case of systems based on FDMA and TDMA methods, systems based on the spread spectrum CDMA technique require a different view of what constitutes a radio channel in this context.

In each cell, one radio channel is set aside for carrying signaling information between the network (i.e., the base station) and the mobile stations in that cell. The signaling channel is used in the mobile-to-BS direction to carry signals for location updating, mobile-originated call setup, and responses to incoming call setup messages (e.g., paging response). In the reverse direction (BS-to-mobile) the signaling channel carries messages related to operating parameters (e.g., location area identity, cell identity), call setup (e.g., paging), and location updating.

2.3.2   Location Updating and Call Setup

To deliver an incoming call to a mobile station, the network (i.e., the MSC and the associated location database) must maintain information on the location of a mobile station as it moves through the coverage area. The mobile station monitors the overhead information broadcast by the network on the signaling channel and updates the operating parameters as necessary. It also checks the location information (e.g., location area identity) broadcast by the new cell and, if it differs from the previous cell, the mobile advises the network of its new location, whereupon the network updates its location register(s). The information is then used to route incoming calls to the MSC currently serving the mobile and to determine the paging broadcast area for the mobile.

The exact procedures for mobile-originated and mobile-terminated call setup depend on the technical standard deployed in a particular mobile system. The procedures described here, however, apply in most cases. A mobile user originates a call by keying in the called number and depressing the *send* key. Note that there is no equivalent of dial tone in a cellular system. The mobile transmits an access request on the uplink signaling channel. If the network can process the call, the BS sends a speech channel allocation message, which enables the mobile to lock on the designated speech channel allocated to that cell while the network proceeds to setup the connection to the called party. A terminal validation procedure may also be invoked as part of the originating call setup to ensure that the terminal originating the call is a legitimate terminal.

For a mobile-terminated call, the network first establishes the current location area for the called mobile through signaling between the home location register (HLR) and the visiting location register (VLR). This process allows the call to be routed to the current serving MSC. The serving MSC initiates a paging message over the downlink signaling channel toward cells contained in the appropriate paging area. If the mobile is turned on, it receives the page and sends a page response to its nearest BS on the signaling channel. The BS sends a speech channel allocation message to the mobile station and informs the network so that the two halves of the connection can be completed.

2.3.3   Handoff and Power Control

During a call, the serving BS monitors the signal quality/strength (C/I ratio) from the mobile. If the signal quality/strength falls below a predesignated threshold, the network requests the neighboring base stations to measure the signal quality from the mobile. If another BS indicates better signal quality/ strength than the serving BS, a signaling message is sent to the mobile on the speech channel (using a *blank-and-burst* procedure) from the current BS asking the mobile to retune to a free channel in the neighboring cell. The mobile retunes to the new channel (in the new cell), and simultaneously the network switches the call to the new BS. Signal quality measurements and new cell selection generally take several seconds, but the change of speech channels (handoff) is essentially transparent to the user except for a very brief break in transmission in FDMA- or TDMA-based systems. A typical intercell handoff situation is shown in Figure 2.4.

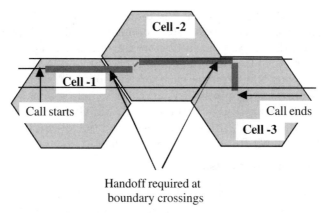

Handoff required at
boundary crossings

Figure 2.4 A typical intercell handoff.

Generally the size of cells within a given cellular system may vary from a radius of 1 km (inner city) to more than 20 km (rural). Thus it is not necessary for the mobile station to transmit at full power at all times to maintain satisfactory signal level at the BS. Most cellular standards therefore allow the BS to signal the mobile to operate at one of a series of transmit power levels, depending on the distance between the mobile and the BS antenna. The main reason for such a feature is to minimize cochannel interference and to conserve terminal battery power.

2.4   INITIAL IMPLEMENTATIONS OF THE CELLULAR CONCEPT: ANALOG CELLULAR SYSTEMS

Though the first-generation (analog) cellular systems are all based on the cellular concept, they were developed at different times in different countries, and were subject to different constraints such as frequency band, channel spacing, and channel coding options. This diversity has led to a number of different analog cellular systems in use in the world today. These *first-generation* cellular systems use analog frequency modulation for speech transmission and frequency shift keying (FSK) for signaling. Thus, individual calls use different frequencies (channels), and the spectrum is shared based on frequency division multiple access (FDMA). More than half the world's cellular subscribers use the AMPS system, which is deployed throughout North and South America as well as in Australia and many countries in Southeast Asia and the Middle East.

The two other prominent (European) systems are Nordic Mobile Telephone (NMT) and the Total Access Communication System (TACS). Whereas the AMPS and TACS systems operate in the 800 and 900 MHz bands, respectively, NMT systems utilize both 450 and 900 MHz bands.

Before digital cellular systems based on TDMA and CDMA technology came on the scene, the AMPS system dominated cellular implementations around the world. For example, just before the introduction of digital cellular systems, over 60% of the subscribers were served by the AMPS system globally. The TACS and NMT systems each served about 13% subscribers, and the remaining were served by other systems. The other systems include C-450 (Germany and Portugal), R2000 (France), RTMS (Italy), NTT (Japan), and JTAC (Japan). Sections 2.4.1 to 2.4.4 provide brief descriptions of AMPS, TACS, NMTS, and NTT; Table 2.1 gives their main characteristics.

2.4.1   The AMPS System

The AMPS system was developed by Bell Laboratories in the mid-1970s, and the first AMPS system was built and tested in Chicago in 1977–78. Cellular mobile service licenses for an initial 40 MHz spectrum in the 800 MHz frequency band were issued, and subsequently another 10 MHz was added. Thus the current spectrum allocation for cellular mobile radio in the United States is 50 MHz within the 824–849 MHz (uplink) and 869–894 MHz (downlink) frequency bands. In a given licensing area, the spectrum is shared by two operators: the wireline common carrier (WCC) and the radio common carrier (RCC). The former is an *arm's-length* subsidiary of a local exchange carrier (LEC), which provides local wired telephone service in the licensing area. Under the AMPS standard, which specifies a carrier spacing of 30 kHz, the 50 MHz spectrum leads to a total of 832 full duplex channels with 416 channels each for the A-band (RCCs) and the B-band (WCCs) operators in each licensing area. Out of these 416 channels, 21 channels are used as control or setup channels, with the remaining 395 channels for user traffic. The latter are then grouped, and each group is assigned to cells within a cluster.

The North American AMPS system uses frequency modulation with 12 kHz deviation for speech (8 kHz for signaling). By adopting the wider 12 kHz deviation for a 30 kHz channel spacing, the AMPS system provides an extended dynamic range for speech and therefore increased protection against cochannel interference. This measure, combined with the use of speech companders, yields a high quality voice channel with the capability to maintain good cochannel performance in a high capacity configuration.

Signaling between the mobile station and the base station is at 10 kb/s; with Manchester encoding, the bit rate is extended to 20 kb/s. The data on the signaling channel are modulated on the radio carrier using direct frequency shift keying with a peak frequency deviation of 8 kHz. Bose–Chaudhary (BCH) encoding is used to protect signaling data from multipath fading. Besides the data transmission on the signaling channel for call setup, data also are transferred on the speech or voice channel: a *blank-and-burst* technique is used, whereby the voice signal is blanked for about 50 ms and a data burst of

Table 2.1 Comparison of the AMPS, TACS, NMT, and NTT Systems

Parameters	AMPS	TACS	NMT 900	NTT
Frequency uplink	824–849 MHz	890–905 MHz	890–905 MHz	860–885 MHz 843–846 MHz
Downlink	869–894 MHz	935–960 MHz	935–960 MHz	15–940 MHz 898–901 MHz
Frequency separation	45 MHz (Rx-to-Tx)	45 MHz (Rx-to-Tx)	45 MHz (Rx-to-Tx)	55 MHz (Rx-to-Tx)
Channel spacing	30 kHz	25 kHz	25/12.5 kHz	25/12.5/6.25 kHz
Number of channels	832 full duplex	600 ful duplex	1999 full duplex	600–2400 full duplex
Voice transmission	FM ±8 kHz deviation	FM ±9.5 kHz deviation	PM ±5kHz deviation	FM ±5 kHz deviation
Data transmission	FSK 10 kb/s[a] ±8 kHz deviation	FSK 8 kb/s[a] ±6.4 kHz deviation	FFSK 1.2 kb/s[b] ±3.5 kHz deviation	FFSK 0.3 kb/s[a] ±4.5 kHz deviation
Error protection code	BCH	BCH	Convolutional code	BCH
Mobile Tx power	3 W nominal	7 W nominal	6 W nominal	5 W nominal
Base station ERP	100 W/channel (max)	100 W/channel (max)	100 W/channel (max)	100 W/channel (max)

[a] Manchester coded.
[b] NRZ coded.

10 kb/s is inserted in the voice channel. This signaling is used for such features as alerting the mobile about an impending channel transfer for a handoff.

To meet the cochannel interference objectives, the typical frequency reuse plan employed in AMPS systems is either a 12-group frequency cluster with omnidirectional antennas or a 7-group cluster with three (120°) sectors per cell. Further, while the call is in progress, the base station transmits a low level supervisory audio tone (SAT) in the region of 6 kHz. Three different SAT frequencies are used by the network and are allocated to the base stations such that base stations most likely to cause interference have a different SAT from the serving base station. The mobile continuously monitors the received SAT and also transponds the signal back to the serving base station. If the mobile (or the base station) detects a difference between the received SAT and the one expected, the audio path is muted to prevent the interfering signal from being overheard. If the condition persists, the call may be aborted.

A narrowband version of AMPS, called NAMPS, which utilizes 10 kHz channels (instead of 30 kHz) has also been standardized by the TIA. The motivation for the NAMPS was to provide a threefold increase in capacity. A number of systems based on the NAMPS standard have been implemented and are in commercial operation.

2.4.2   The TACS System

The TACS system developed in the United Kingdom is an adaptation of the North American system, AMPS, to suit European frequency allocations. It utilizes the 50 MHz spectrum in the 890–915 MHz (uplink) and 935–960 MHz (downlink) range allocated by the 1979 World Radio Administrative Council (WARC 79) for mobile radio service. With the 25 kHz channel spacing specified by the Conference of European Postal and Telecommunication Administrations (CEPT), TACS provides 1000 full duplex channels. The frequency deviation for speech is ±9.5 kHz peak and a frequency deviation of ±6.4 kHz for data transmission at 8 kb/s rate. Only the first 600 channels were allocated to the two analog mobile network operators in the United Kingdom; the remaining top 400 channels (2×10 MHz spectrum) were held in reserve for the second-generation digital cellular system (GSM), which is now fully operational.

TACS essentially retains the AMPS signaling scheme with some enhancements to location registration procedures to provide seamless, countrywide roaming capabilities. TACS also provides some extra features like charge information for pay phones (e.g., in taxis). The TACS system has been adopted elsewhere in Europe (Ireland, Spain, and Austria) as well as in the Middle East (Kuwait, the United Arab Emirates, and Bahrain) and Far East (Hong Kong, Singapore, Malaysia, and China). JTACS, a variant of TACS, is deployed in some regions in Japan.

2.4.3   The NMT System

The Nordic Mobile Telephone (NMT) system was designed, developed, and implemented in (Denmark, Norway, Sweden, and Finland) in 1981, using the 450 MHz band (453–457.5 and 463–467.5 MHz). To alleviate capacity problems of NMT450, it was upgraded to NMT900 in 1986 to occupy 50 MHz in the 890–915 and 935–960 MHz band. The NMT450 system was designed to provide full roaming capability within the Nordic countries.

NMT450 uses a channel spacing of 25 kHz, speech modulation being analog FM with frequency deviation of ± 5 kHz. NMT900 also uses ± 5 kHz deviation, but the channel spacing is only 12.5 kHz, which doubles the available number of channels (but also degrades adjacent channel rejection performance and complicates the radio design). Signaling is at 1.2 kb/s using, fast frequency shift keying (FFSK) with convolution forward error correcting code for error protection.

Besides the Nordic countries, NMT450 has been deployed in Austria, Spain, Netherlands, Belgium, France, Iceland, Turkey, and Hungary. However, roaming is not possible between all the NMT450 networks because of differences in frequency allocations between countries. NMT900 has been deployed in many countries as an overlay network to overcome capacity limits reached on the NMT450.

2.4.4   The NTT System

The first NTT cellular system in the 800 MHz band was introduced in Japan (Tokyo) in 1979 with a 30 MHz allocation in the 925–940 MHz (uplink) and 870–885 MHz (downlink) bands with 25 kHz channel spacing (600 channels) and a control channel signaling rate of 300 b/s. The current frequency allocations provide for 56 MHz in the 860–885/915–940 and 843–846/898–901 frequency bands with a channel spacing of 25 kHz.

The high capacity system introduced in 1988 uses a reduced channel spacing of 12.5 kHz and an increased control channel signaling rate of 2400 b/s. The number of radio channels was further increased to 2400 by frequency interleaving, which can provide a nominal spacing of 6.25 kHz. In addition, the high capacity system uses a 100 b/s associated control channel (under the voice channel) that permits signaling without interrupting speech traffic. Nationwide coverage and roaming is facilitated through the use of dual-mode mobile terminals that can support the channel structures based on 25 and 12.5 kHz frequency spacing. The cellular mobile service was deregulated in Japan in 1987, and three cellular service providers (NTT DoCoMo, IDO, and DDI Cellular) are licensed to provide cellular service in each of the 10 licensing regions in Japan.

2.5 CONCLUDING REMARKS

The defining events for today's mobile communications services were the development of the cellular concept and the introduction of mobile communication systems based on this concept. Evolution of these analog cellular mobile systems toward the support of small, affordable, handheld terminals, and wide-area roaming capabilities caught the attention of average consumers, leading to unprecedented growth in mobile telephony services. In spite of the introduction of digital cellular services in early 1990s to provide higher capacity and better performance, analog cellular systems still dominate the cellular mobile market in terms of total subscribers—though their rate of implementation is now rapidly declining in favor of the second-generation digital cellular systems.

2.6 REFERENCES

[01] V. H. Macdonald, "The Cellular Concept," *Bell System Technical Journal,* Vol. 58, No. 1, January 1979.

[02] A. Mehrotra, *Cellular Radio: Analog and Digital Systems,* Artech House, Boston, 1994.

[03] M. Appleby and F. Harrison, "Cellular Radio Systems," Chapter 47, in *Telecommunication Engineer's Handbook.* F. Mazda, ed., Butterworth, Heinmann, London, 1995.

[04] TIA IS-553, "Cellular System Mobile Station–Land Station Compatibility Specifications," Telecommunications Industries Association, Interim Standard #553, TIA, Washington, DC, 1989.

[05] TIA IS-88, "Mobile Station–Land Station Compatibility Standard for Dual-Mode Narrowband Analog Cellular System," Telecommunication Industries Association Interim Standard #88, TIA, Washington, DC, 1993.

[06] K. Watanabe et al., "High Capacity Land Mobile Communication Systems in NTT," IEEE Vehicular Technology Conference, Tampa, FL, 1987.

[07] Nordic Mobile Telecommunication System Description, NMT Doc 1, Oslo, 1977.

CHAPTER 3

DIGITAL CELLULAR MOBILE SYSTEMS

3.1 INTRODUCTION

The analog mobile systems were originally targeted for a relatively select group of users who mostly had the mobile telephones installed in their vehicles. However, the demand for cellular mobile services has increased dramatically with the availability of low cost, small, lightweight, handheld mobile terminals that have relatively long battery life and can be used in a variety of environments—for example, on the street and inside buildings, as well as in vehicles. Theoretically, cellular systems can provide unlimited capacity through addition of new cells and splitting and/or sectorization of existing cells. Because of the nature of the interference characteristics of analog signals and the relatively limited analog technology options to combat interference effects, current analog cellular systems are unable to accommodate the forecasted demand for mobile services.

Concepts of digital radio were first deployed in military applications to provide improved reception in a highly interference-prone environment and to provide a high level of security (through encryption) against eavesdropping on the radio path and against unauthorized access (through strong authentication and verification procedures). The driving force behind commercial digital cellular systems is not only the increased system capacity but the reduction of mobile terminal size and the average power requirements, which in turn increases the terminal battery life and reduces the terminal cost. The development of low rate codecs, the dramatic increase in the device densities of integrated circuits, and the advances in digital signal processing (DSP) have made completely digital (second-generation) cellular mobile systems a commercial reality. The advantages of digital cellular systems can be summarized as follows:

- three-to tenfold capacity increase over analog systems
- reduced RF transmission power and longer battery life

- international and wide-area roaming capability

- better security against fraud (through terminal validation and user authentication)

- encryption capability for information security and privacy

- compatibility with ISDN, leading to wider range of services

- ability to operate in small (micro) cell environments

Spectrum sharing in the digital environment can be based on either time division multiple access (TDMA) or code division multiple access (CDMA). In the former case each radio channel is partitioned into a number of time slots, and each mobile user is assigned a frequency/time slot combination on a demand basis. In case of CDMA, a radio channel can be used simultaneously by multiple mobile users, and the signals from different users are distinguished by spreading them with different *pseudo–random noise* (PN) codes.

Currently three digital cellular standards are based on TDMA technology, one for each major economic region. The European standard is GSM (global system for mobile communications) which is very comprehensive and is intended to replace the current analog systems in Europe. The GSM system and its frequency-upshifted version (DCS1800) has been adopted by a large number of operators worldwide and has captured the largest global subscriber base among the current digital cellular mobile systems.

The TDMA-based digital cellular standard (TIA IS-54) in the United States is designed to coexist with the analog (AMPS) system using dual-mode terminals, with intersystem operations (roaming) provided by the TIA IS-41 standard (now adopted by the American National Standards Institute as the ANSI-41 standard). Cellular mobile systems based on this standard are being implemented in North and South America, where they coexist with AMPS analog systems. An enhanced version of IS-54 (called TIA IS-136), which utilizes a digital control channel and offers many additional capabilities, has also been standardized. Based on this enhancement, the North American TDMA system is now generally identified as IS-136. In the absence of a definitive name for this system, it is referred to as D-AMPS (digital-AMPS) in this book.

The Japanese have developed the personal digital cellular (PDC) standard, which is similar to the D-AMPS. The implementation of systems based on the PDC is well under way in Japan. In fact, according to recent forecasts, it is expected that the PDC system will saturate soon after year 2000, which is one of the reasons for the Japanese interest in and drive for the third-generation mobile communications standards in the ITU.

A CDMA-based cellular mobile system has been standardized in the United States as TIA IS-95. CDMA cellular systems claim to provide a significant capacity and cost advantage over analog- and TDMA-based systems.

CDMA systems based on the IS-95 standard are now in commercial operation in North and South America as well as in Japan, Korea, and China.

Second-generation cellular mobile systems using digital technology are now being deployed either in parallel with or as replacement of existing analog cellular systems. The sections that follow describe in some detail the major digital cellular mobile systems being implemented.

3.2   GSM: THE EUROPEAN TDMA DIGITAL CELLULAR STANDARD

3.2.1   GSM Standardization and Service Aspects

The GSM standard was developed by the Groupe SpecialMobile, which was an initiative of the Conference of European Post and Telecommunications (CEPT) administrations. The underlying aim was to design a uniform pan-European mobile system to replace the existing incompatible analog systems. Work on the standard was started in 1982, and the first full set of specifications (phase 1) became available in 1990. The responsibility for GSM standardization now resides with the Special Mobile Group (SMG) under the European Telecommunication Standards Institute (ETSI), and revisions/enhancements to various aspects of GSM standard are being carried out in SMG technical subcommittees. Figure 3.1 shows the structure of SMG and the primary responsibilities of the various SMG subcommittees.

The characteristics of the initial GSM standard include the following:

- fully digital system utilizing the 900 MHz frequency band
- TDMA over radio carriers (200 kHz carrier spacing)
- 8 full-rate or 16 half-rate TDMA channels per carrier
- user/terminal authentication for fraud control
- encryption of speech and data transmissions over the radio path
- full international roaming capability
- low speed data services (up to 9.6 kb/s)
- compatibility with ISDN for supplementary services
- support of short message service (SMS)

GSM supports a range of basic and supplementary services, and these services are defined in terms analogous to those for ISDN (i.e., bearer services, teleservices, and supplementary services). The most important service supported by GSM is telephony. Other services derived from telephony included in the GSM specification are *emergency calling* and *voice messaging*. Emergency call-

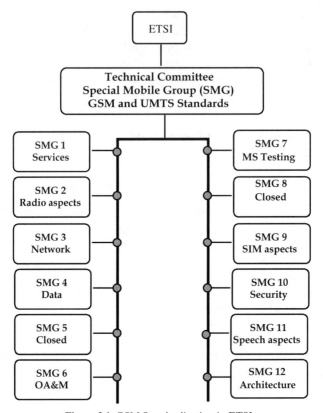

Figure 3.1 GSM Standardization in ETSI.

ing refers to such services as E.911 in North America and E.112 in Europe. Bearer services supported in GSM include various asynchronous and synchronous data services for information transfer between GSM and other networks (i.e., PSTN, ISDN, CSPDN, PSPDN) at rates from 300 to 9600 b/s. Teleservices based on these bearer services include group 3 fax and short message service (SMS). The data capabilities of GSM have now been enhanced to include high speed circuit-switched data (HSCSD) and general packet radio service (GPRS). The common ISDN-like supplementary services supported by GSM include the following:

- call offering services—call forwarding
- call restriction services—call barring
- call waiting service

- call hold service
- multi party service—tele conferencing
- calling line presentation restriction services
- advice of charge service
- closed user group service

Whereas the first two categories of supplementary services (call offering and call restriction) are defined in the original GSM specification (phase 1), the remaining services were recently added as part of enhancements to GSM (phase 2). Many of these services are similar (at least in their description) to services currently available in fixed networks. However, because of the mobile radio link, their implementation may impose special requirements. Mobile data communication services provided by cellular systems are addressed in Chapter 5.

The GSM standard has been undergoing continuous extensions and enhancements to support more services and capabilities like high speed circuit-switched data (HSCSD), general packet radio service (GPRS), and CAMEL (customized applications for mobile network-enhanced logic).

An 1800 MHz version of GSM, known as DCS1800, has also been standardized in Europe. Further, Committee T1P1 and TIA TR46 are responsible for the PCS1900 standard in the United States, which is based on the GSM and DCS1800 European standards. A number of PCS1900 systems are now operational and they represent North American implementations of GSM.

3.2.2   GSM Reference Architecture and Function Partitioning

The reference architecture and associated signaling interfaces for GSM are closely aligned with those specified in ITU-T Recommendation Q.1001. As shown in Figure 3.2, the GSM system comprises base transceiver stations (BTS), base station controllers (BSC), mobile switching centers (MSC), and a set of registers (databases) to assist in mobility management and security functions. All signaling between the MSC and the various registers (databases) as well as between the MSCs takes place using the Signaling System 7 (SS7) network, with the application level messages using the Mobile Application Protocol (MAP) designed specifically for GSM. The MAP protocol utilizes the lower layer functions (TCAP, SCCP, MTP) from the SS7 protocol stack. The signaling between the MSC and the national PSTN/ISDN is based on the national options for SS7 telephone user part (TUP) or ISDN user part (ISUP). The base signaling transport for the A interface is also SS7. The interface specifications for GSM are addressed in greater detail in a later section.

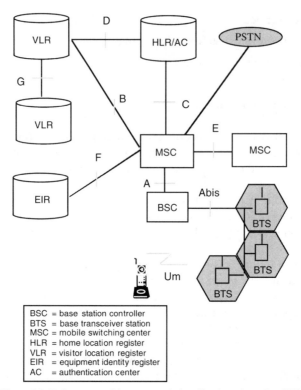

Figure 3.2 Reference architecture and signaling interfaces for GSM.

3.2.2.1   Mobile Station (MS).   The GSM mobile stations are portable radiotelephony units that can be used on any GSM system as vehicular and/or handheld terminals. Power levels supported by the GSM mobile station currently range from 0.8 to 8.0 W, and power-saving techniques are used on the air interface to extend battery life. At the time of manufacture, an international mobile equipment identity (IMEI), which is not easily alterable or removable, is programmed into the terminal.

A subscriber identity module (SIM) is required to activate and operate a GSM terminal. The SIM may be contained within the MS, or it may be a removable unit that can be inserted by the user. In the latter case, any GSM terminal (capable of receiving a detachable SIM card) can become the user's MS upon plugging in the SIM card. The international mobile subscriber identity (IMSI) is programmed into the SIM at the time of service provisioning, along with the appropriate security parameters and algorithms.

3.2.2.2   Base Station System (BSS).   The base station system comprises a base station controller (BSC) and one or more subtending base transceiver stations (BTS). The BSS is responsible for all functions related to the radio resource (channel) management. This includes the management of radio channel configuration with respect to use as speech, data, or signaling channels, allocation and release of channels for call setup and release; control of frequency hopping, and transmit power at the mobile station (MS). The range of functions performed by the BSS therefore include the following:

- Radio resource control
    configuration of radio channels
    selection, allocation, and deallocation of radio channels
    monitoring of radio channel busy/idle status
    encryption of radio interface
- Frequency hopping and power control
    assignment of frequency-hop sequence and start time
    assignment of effective radiated power (ERP) values to mobile stations
- Handoff management
    collect signal quality data from adjacent BSSs
    analyze signal quality data and determine handoff need
    keep MSC informed regarding handoff activity
- Digital signal processing
    transcoding and rate adaption
    channel coding and decoding

As mentioned earlier, the BSS functions are partitioned into a BSC and a BTS (a single BSC generally controlling multiple BTSs). Table 3.1 lists some of the functions and/or physical elements located in the BSC and BTS.

Table 3.1 Functions and/or Physical Elements Located in a BSC and a BTS

Base Station Controller (BSC)	Base Transceiver Station (BTS)
BSC processor	BTS processor
Control of BTS(s)	Radio transceivers
Radio resource management	Equalizers
Handoff management and control	Channel coders
	Transcoders
	Encryption unit
	Interleaving and deinterleaving
	Combiners
	Preselectors

3.2.2.3 *Mobile Switching Center (MSC).* The mobile switching center for GSM can be viewed as a local ISDN switch with additional capabilities to support mobility management functions like terminal registration, location updating, and handoff. Further, unlike a local switch in a fixed network, the MSC does not contain the mobile subscriber parameters [which are dynamic and are held in the visitor location register (VLR) associated with the MSC]. Thus, the MSC performs the following major functions:

- Call setup, supervision, and release
- Digit collection and translation
- Call routing
- Billing information collection
- Mobility management
  registration
  location updating
  inter-BSS and inter-MSC call handoffs
- Paging and alerting
- Management of radio resources during a call
- Echo cancellation
- Manage connections to BSS, other MSCs, and PSTN/ISDN
- Interrogation of appropriate registers (V/HLRs)

3.2.2.4 *Home Location Register (HLR).* The HLR represents a centralized database that has the permanent datafill about the mobile subscribers in a large service area (generally one per GSM network operator). It is referenced using the SS7 signaling capabilities for every incoming call to the GSM network for determining the current location of the subscriber [i.e., for obtaining the mobile station routing number (MSRN) so that the call may be routed to the mobile station's serving MSC]. The HLR is kept updated with the current locations of all its mobile subscribers, including those who may have roamed to another network operator within or outside the country. The routing information is obtained from the serving VLR on a call-by-call basis, so that for each incoming call the HLR queries the serving VLR for an MSRN.

Usually one HLR is deployed for each GSM network for administration of subscriber configuration and service. Besides the up-to-date location information for each subscriber, which is dynamic, the HLR maintains the following subscriber data on a permanent basis:

- international mobile subscriber identity (IMSI)
- service subscription information

- service restrictions
- supplementary services (subscribed to)
- mobile terminal characteristics
- billing/accounting information

3.2.2.5   Visiting Location Register (VLR).   The VLR represents a temporary data store, and generally there is one VLR per MSC. This register contains information about the mobile subscribers who are currently in the service area covered by the MSC/VLR. The VLR also contains information about locally activated features such as call forward on busy. Thus, the temporary subscriber information resident in a VLR includes:

- features currently activated
- temporary mobile station identity (TMSI)
- current location information about the MS (e.g., location area and cell identities)

3.2.2.6   Authentication Center AC).   Generally associated with the HLR, the authentication center contains authentication parameters that are used on initial location registration, subsequent location updates, and on each call setup request from the MS. In case of GSM, the AC maintains the authentication keys and algorithms, and provides the security triplets (RAND, SRES, and Kc) to the VLR so that the user authentication and radio channel encryption procedures may be carried out within the visited network. The authentication center for GSM contains the security modules for the authentication keys (Ki) and the authentication and cipher key generation algorithms A3 and A8, respectively.

3.2.2.7   Equipment Identity Register (EIR).   The EIR maintains information to authenticate terminal equipment so that fraudulent, stolen, or non-type-approved terminals can be identified and denied service. The information is in the form of white, gray, and black lists that may be consulted by the network when it wishes to confirm the authenticity of the terminal requesting service.

3.2.3   GSM Radio Aspects

3.2.3.1   Basic Radio Characteristics.   In GSM the uplink (mobile-to-base) frequency band is 890–915 MHz and the corresponding downlink (base-to-mobile) band is 935–960 MHz, resulting in a 45 MHz spacing for duplex operation. The GSM uses time division multiple access (TDMA) and frequency division multiple access (FDMA), whereby the available 25 MHz spectrum is partitioned into 124 carriers (carrier spacing = 200 kHz), and each car-

Table 3.2 Radio Parameters and Characteristics for GSM

System Parameter	Value (GSM)
Multiple access	TDMA/FDMA/FDD
Uplink frequency (mobile-to-base)	890–915 MHz
Downlink frequency (base-to-mobile)	935–960 MHz
Channel bandwidth	200 kHz
Number of channels	124
Channels/carrier	8 (full rate), 16 (half rate)
Frame duration	4.6 ms
Interleaving duration	40 ms
Modulation	GMSK
Speech coding method	RPE-LTE convolutional
Speech coder bit rate	13 kb/s (full rate)
Associated control channel	Extra frame
Handoff scheme	Mobile-assisted
Mobile station power levels	0.8, 2, 5, 8 W

rier in turn is divided into 8 time slots (radio channels). Each user transmits periodically in every eighth time slot in an uplink radio carrier and receives in a corresponding time slot on the downlink carrier. Thus several conversations can take place simultaneously at the same pair of transmit/receive radio frequencies. The radio parameters for GSM are summarized in Table 3.2; for the frame structure see Figure 3.3.

In the GSM system, digitized speech is passed at 64 kb/s through a speech coder (transcoder), which compresses the 64 kb/s PCM speech to a 13 kb/s data rate. The transcoder models the vocal tract of the user and generates a set of filter parameters that are used to represent a segment of speech (20 ms long), and only the filter parameters and the impulse input to the filter are transmitted on the radio interface. The speech coding technique improves the spectral efficiency of the radio interface, thereby increasing the traffic capacity of the system (more users over a limited bandwidth). The linear predictive, low bit rate (LBR) transcoder is based on residual pulse excitation–long-term prediction (RPE-LTE) techniques. The GSM transcoder also permits the detection of silent periods in the speech sample, during which transmit power at the mobile station can be turned off to save power and extend battery life.

The transcoded speech is error protected by passing it through a channel encoder, which utilizes both a parity code and a convolution code. The channel encoding increases the bit rate of transcoded speech from 13 kb/s (260

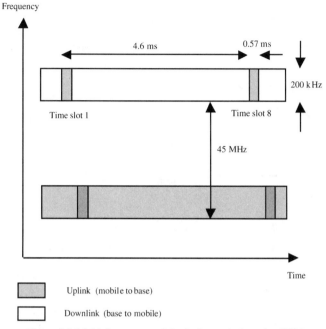

Figure 3.3 Multiple access and duplexing technique for GSM.

bits/20 ms of speech) to 22.8 kb/s (456 bits/20 ms of speech) for the GSM full-rate coder (22.8 kb/s, 8 slots/frame). A half-rate coder for GSM (11.4 kb/s, 16 slots/frame) has also been specified. The 456 bits of convolutionally encoded data (representing 20 ms speech sample) is then interleaved to combat the effects of burst errors on the radio path. An interleaving depth of 8 is used in GSM for full rate speech, which requires that the channel encoded bits produced over two adjacent 20 ms intervals (2 × 456 bits) be split into eight blocks (114 bits each) and transmitted over eight frames. The interleaved data are then modulated by means of Gaussian minimum shift keying (GMSK) and passed through a duplexer, which provides filtering to isolate transmit and receive signals. This process of information transfer for speech signals, in the uplink direction across the radio interface, is illustrated in Figure 3.4. A similar process is applicable for the downlink.

Whereas the digital signal processing in GSM radio just described leads to higher capacity and better speech signal quality, it introduces additional delays. For example, the delays introduced by speech coding (20 ms), interleaving (37 ms), analog-to-digital (A/D) conversion (8 ms), and processing for transmission and switching (15 ms) add up to 80 ms. For a connection into/from the PSTN, this 80 ms delay will be in addition to the existing PSTN

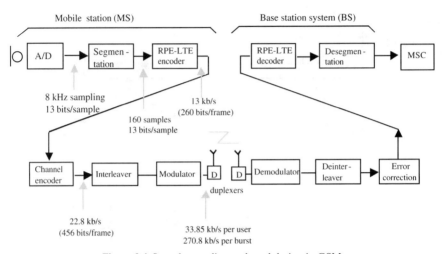

Figure 3.4 Speech encoding and modulation in GSM.

delay. Therefore echo control devices become necessary for all GSM calls that use the PSTN as a transit network. To remove any echoes being returned from the PSTN to the GSM mobile user, the PSTN interface of the GSM network includes an echo canceller.

3.2.3.2   Logical Channels.   There are two distinct categories of logical channels in GSM: traffic channels (TCH) and control channels (CCH). In the initial GSM specification only the full-rate speech channels (22.8 kb/s) and data channels (9.6, 4.8, and 2.4 kb/s) are defined as TCH. The three types of control channel are broadcast (BCCH), common (CCCH), and dedicated (DCCH), each of which is further subdivided. Figures 3.5 presents the various subcategories of traffic and control channels in GSM.

Each traffic channel type shown in Figure 3.5 is always associated with a control channel (associated control channel). Currently only the full-rate speech channel and the data channels have been fully defined and implemented; the definition of half-rate channel has been completed and is ready for implementation.

The broadcast control channel (BCCH) is a unidirectional (base-to-mobile) channel that is used to broadcast (continuous transmission) information regarding the mobile's serving cell as well as neighboring cells. It may include a frequency correction channel (FCCH) and a synchronization channel (SCCH) to provide accurate tuning to BS and frame synchronization, respectively.

A common control channel (CCCH) may be used either for uplink (mobile-to-base) or downlink (base-to-mobile) communications. The pag-

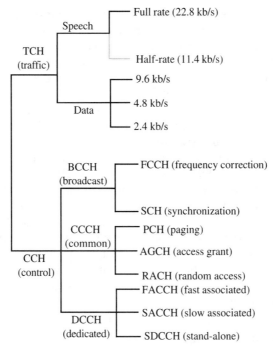

Figure 3.5 Logical channel structure in GSM.

ing (PCH) and access grant (AGCH) channels operate in the downlink direction—the former for paging a mobile and the latter to assign dedicated resources to the mobile (i.e., SDCCH). When in the idle mode, the MS always listens to the paging channel for incoming calls.

The dedicated control channels (DCCH) are used for call setup, or for measurement and handoff, and are assigned to a single mobile connection. The slow associated control channel (SACCH) is either assigned to a traffic channel (TCH) or to a stand-alone dedicated control channel (SDCCH). The SACCH is deployed to transfer link quality and signal strength information in the uplink direction and power control frame adjustment (timing) information in the downlink direction. The fast associated control channel (FACCH) is assigned only to a traffic channel (TCH) and relies on frame stealing to transfer handoff information during an active call. The SDCCH provides data transfer in both the uplink and downlink directions for call setup or short messages and is released at the completion of data transfer.

3.2.3.3   *Frequency Hopping and Discontinuous Transmission.* To reduce the effects of multipath fading when a mobile station is stationary or moving slowly, GSM provides frequency-hopping capability on the radio interface. Frequency hopping also provides added security against unauthorized eavesdropping on a call in progress. Frequency hopping is achieved in such a way that, according to the calculated sequence (based on a simple algorithm) on both sides of the air interface, the MS sends and receives each time slot (burst) on a different frequency. Frequency hopping is applied only on the traffic and nonbroadcast channels.

GSM utilizes voice-activated transmissions or discontinuous transmissions (DTx) to maximize spectrum efficiency. The technique is based on detecting voice activity and switching on the transmitter only during periods when there is active speech to transmit. Further, switching off the transmitter during silent periods reduces interference in the air, thereby allowing the use of smaller frequency reuse clusters, and reduces battery power consumption in the mobile terminal. Up to 50% gain in spectrum efficency can be achieved through DTx. An adaptive threshold voice activity detector (VAD) algorithm is used in the GSM radio equipment to detect silent periods and interrupt transmission. At the receiving end, the DTx is detected and empty frames are filled with comfort noise. The VAD algorithm and its implementation need to be carefully designed to minimize speech clipping and the resulting degradation in speech quality.

3.2.4 Security Aspects

The security procedures in GSM are aimed at protecting the network against unauthorized (fraudulent) access and protecting the privacy of the mobile subscribers against eavesdropping on their communications. The security procedures also prevent unauthorized parties from tracing the identity and location of the subscribers as they roam within or outside the home network. In GSM, protection from unauthorized access is achieved through strong authentication procedures that validate the true identity of the subscriber before he or she is permitted to receive service. Eavesdropping on subscribers' communications is prevented by ciphering the information channel across the radio interface (i.e., applying encryption on the digital stream on the radio path). To protect the identity and location of the subscriber, the appropriate radio signaling (control) channels are also ciphered, and a temporary subscriber identity (TMSI) instead of the actual identity (IMSI) (is used over the radio path). Note that the privacy mechanisms (encryption and use of TMSI) are used only over the radio path, not within the fixed infrastructure, where the communications are transmitted in the clear, as they are in PSTN/ISDN.

In GSM systems, each mobile user is provided with a subscriber identity module (SIM). Two possible versions of SIM are defined in GSM standard: one version is a chip card the size of a credit card to be inserted in the mobile terminal; the other version is a small (25 mm $\times$ 15 mm) plug-in SIM that can be installed in the mobile terminal on a semipermanent basis. The latter is specially suited to small GSM handsets. At the time the mobile terminal is powered on, the subscriber may be required to enter a four- to eight-digit personal identification number (PIN) to validate the ownership of the SIM. Until the PIN has been verified, the MS will not be able to receive service or access the personal data held in the SIM.

At the time of service provisioning the IMSI, the individual subscriber authentication key (Ki), the authentication algorithm (A3), the cipher key generation algorithm (A8), and the encryption algorithm (A5) are programmed into the SIM by the GSM operator. The IMSI and the secret authentication key (Ki) are specific to each MS; the authentication algorithms (A3) and the cipher key generation algorithm (A8) can be different for different network operators; the encryption algorithm (A5) is unique and needs to be used across all GSM network operators.

The authentication center is responsible for all security aspects, and its function is closely linked with the HLR. The AC generates the Ki's, associates them with IMSIs, and provides for each IMSI a set of triplets consisting of RAND (random number), SRES (signed **res**ponse), and Kc (ciphering key). The HLR then provides the appropriate VLR with this set, and it is the VLR that carries out the authentication check and provides the appropriate ciphering key (Kc) to the BTS for encryption/decryption of the radio path. It is also possible for the new VLR to receive unused triplets from the old VLR at location update. Further, the serving VLR can request additional triplets from the HLR/AC if the current set is depleted below a certain threshold. The network operator has the option of invoking the procedure at one or more of the following instances: initial registration, location update, and call origination/termination. Figure 3.6 summarizes the general authentication process and ciphering key generation in GSM.

Before the network can authenticate the user and ciphering can begin on the radio path, the network needs to know the identity (IMSI) associated with the MS. However, passing the IMSI on a clear channel (e.g., at initial registration or location update) can compromise security in terms of protecting the identity and location of the subscriber. To alleviate this potential security threat, GSM uses an identity alias, the temporary mobile subscriber identity (TMSI), which is used instead of IMSI, wherever possible. TMSI associated with an MS thus has a one-to-one relationship with its IMSI and is always used over the radio path when the MS is in a given location area. When the MS moves to a new location area, initially the TMSI and the location area identity (LAI) of the previous location area are sent over the radio path so that the new VLR can interrogate the old VLR for the true identity (i.e., IMSI) of the MS. The new VLR then assigns a new TMSI (and cancels the old one). Since the TMSI can be coded in four octets (as opposed to nine octets for IMSI), use of TMSI also reduces the signaling load on the radio channel.

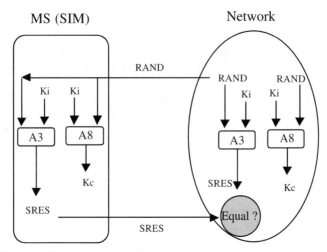

Figure 3.6 General authentication process and ciphering key generation.

An additional security feature in GSM is the equipment identity register (EIR), which maintains black, gray, and white lists of international mobile equipment identities (IMEIs) for monitoring mobile equipment. Each mobile terminal is assigned a unique IMEI, which consists of a type approval code (TAC), a final assembly code (FAC), and a serial number. The IMEI is used to validate mobile equipment (terminals) so that non-type-approved, faulty, or stolen terminals are denied service. Appendix A describes the structure of IMSI, IMEI, and other numbers and identities.

3.2.5 GSM Protocol Model

As shown in Figure 3.7 the signaling at the radio interface (Um) consists of LAPDm at layer 2. LAPDm is a modified version of LAPD (link access protocol for D channel), used in ISDN user–network access, to accommodate radio interface specific features. Layer 3 is divided into three sublayers that deal with radio resource management (RR), mobility management (MM), and connection management (CM), respectively. The radio resource management is concerned with managing logical channels, including the assignment of paging channels, signal quality measurement reporting, and handoff. The mobility management sublayer provides functions necessary to support user/terminal mobility, such as terminal registration, terminal location updating, authentication, and IMSI detach/attach. The connection management sublayer is concerned with call and connection control, establishing and clearing calls/connections, management of supplementary services, and support of the short message service.

The Abis interface between the BTS and the BSC, though fully defined in

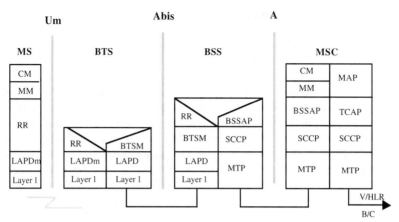

Figure 3.7 Protocol model for GSM.

the GSM standard, tends to be a proprietary interface in most GSM implementations. The radio resource layer (RR′) in the BTS is responsible for channel establishment and release, handoff, and paging. The BTS management (BTSM) layer looks after the management of all aspects of the radio channels, including radio link layer, control channels, and transceiver management.

The BSC-to-MSC interface (A interface) and the interfaces between an MSC and a V/HLR or another MSC deploy ITU-T Signaling System 7 (SS7) using the MTP (Message Transfer Part), SCCP (Signaling Connection Control Part), and TCAP (Transaction Capability Application Part) and the Mobile Application Protocol (MAP). However, the MAP used in GSM is not an ITU-T standard but a GSM-specific protocol developed as part of the GSM standard (i.e., an ETSI standard). Connections between the MSC and other PSTN/ISDN exchanges utilize the Telephone User Part (TUP) or the ISDN User Part (ISUP) specific to the country or region.

3.2.6   Typical Call Flow Sequences in GSM

Simplified signaling sequences for some basic features supported in GSM are described in Sections 3.2.6.1 to 3.2.6.5. These features are not unique to the GSM but need to be supported in all cellular systems. Each individual system (standard), such as GSM, D-AMPS, PDC, or IS-95 CDMA, may support these features in somewhat different ways depending on system architecture, interface standards, authentication and ciphering procedures, and so on. The signaling sequences to be described address the following features:

- location updating
- mobile call origination

- mobile call termination
- authentication and ciphering
- inter-MSC call handoff

3.2.6.1 *Location Updating.*   As users move about within and outside their home service area, the home system must know the location of all active mobile stations in real time in order to deliver incoming calls. The location updating feature is invoked when an active MS moves from one location area to another or when the MS tries to access the network and it is not already registered in the serving VLR for its present location. Location areas generally

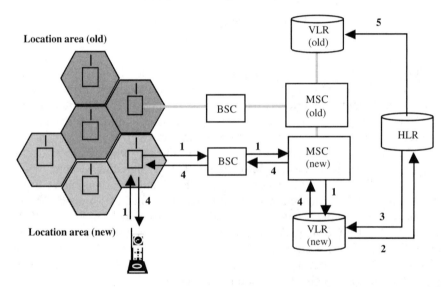

1. The MS sends a Location Update request to the VLR (new) via the BSS and MSC.

2. The VLR sends a Location Update message to the HLR serving the MS which includes the address of the VLR (new) and the IMSI of the MS. This updating of the HLR is not required if the new LA is served by the same VLR as the old LA.

3. The service and security related data for the MS is downloaded to the new VLR.

4. The MS is sent an acknowledgment of successful location update.

5. The HLR requests the old VLR to delete data relating to the relocated MS

Figure 3.8 Location updating sequence in GSM.

consist of multiple, contiguous cells and are identified by location area identities (LAI). In GSM, the MS continuously monitors the transmissions from the surrounding base stations. When it determines that it has moved to a new location area, it initiates the signaling sequence (Figure 3.8), which assumes that the new location area is served by a different BSS, MSC, and VLR.

3.2.6.2   Mobile Call Origination.   Initially when the user enters the called number and presses the *send* key, the MS establishes a signaling connection to the BSS on a radio channel. This may involve authentication and ciphering. Once this has been established, the call setup procedures will take place according to the sequence shown in Figure 3.9.

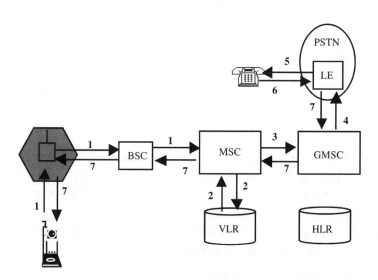

1. The MS sends the dialed number indicating service requested to the MSC (via BSS).

2. The MSC checks from the VLR if the MS is allowed the requested service. If so, MSC asks the BSS to allocate necessary resources for the call.

3. If the call is allowed, the MSC routes the call to GMSC.

4. The GMSC routes the call to the Local Exchange of called user.

5. The LE alerts (applies ringing) the called terminal.

6. Answer back (ring back tone) from the called terminal to LE

7. Answer back signal is routed back to the MS through the serving MSC which also completes the speech path to the MS.

Figure 3.9 Mobile call origination in GSM.

3.2.6.3   Mobile Call Termination.   The sequence shown in Figure 3.10 relates to a call originating in the PSTN and terminating at an MS in a GSM network.

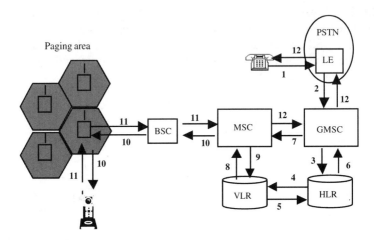

1. The PSTN user diales the MSISDN of the called user in GSM.

2. The LE routes the call to the GMSC of the called GSM user.

3. The GMSC uses the dialed MSISDN to determine the serving HLR for the GSM user and interrogates it to obtain the required routing number.

4. The HLR requests the current serving VLR for the called MS for a MSRN (MS Roaming Number) so that the call can be routed to the correct MSC.

5. The VLR passes the MSRN to the HLR.

6. The HLR passes the MSRN to the GMSC.

7. Using the MSRN, the GMSC routes the call to the serving MSC.

8. The MSC interrrogates the VLR for the current Location Area Indentity (LAI) for the MS.

9. The VLR provides the current location (LAI) for the MS.

10. The MSC pages the MS via the appropriate BSS. The MS responds to the page and sets up the necessary signaling links.

11. When the BSS has established the necessary radio links, the MSC is informed and the call is delivered to the MS.

12. When the MS answers the call, the connection is completed to the caling PSTN user.

Figure 3.10 Call flow for a mobile-terminated call.

3.2.6.4 Authentication and Encryption.   The authentication and ciphering functions in GSM are closely linked and are performed as a single procedure between the MS and the network. As outlined in Section 3.2.4, the security procedure in GSM is based on the so-called private key (or symmetric key) mechanism, which requires that a secret key (called Ki) be allocated and programmed into each mobile station. An authentication algorithm (A3), a cipher key generation algorithm (A8), and an encryption algorithm (A5) are also programmed into the MS at the time of service provisioning. The relevant call flows are shown in Figure 3.11.

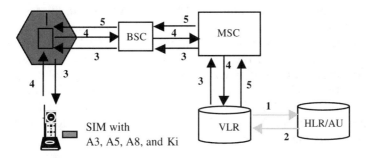

1. At terminal location update, VLR sends IMSI to the HLR.
2. HLR returns security triplets (RAND, SRES, Kc) to the VLR.
3. For authentication and ciphering the VLR sends RAND to the MS.
4. Using stored A3 algorithm and secret key Ki stored in the SIM, and RAND provided by the VLR, the MS calculates the SRES and returns it to the VLR. Using the A8 algorithm and Ki, the MS also calculates the cipher key Kc.
5. If the SRES returned by the MS matches with the stored SRES in the VLR, the VLR sends the cipher key Kc to the BTS which uses Kc for ciphering the radio path (downlink).

The MS uses its Kc to cipher the radio path (uplink) using encryption algorithm A5

Figure 3.11 Call flows for authentication and ciphering.

3.2.6.5 Inter-MSC Handoff.   Cellular systems must be able to provide the capability to hand off *calls in progress* from one channel to another. Handoff of calls already in progress from one channel to another may be invoked for one of the following reasons:

 • to avoid dropped calls when a subscriber (with call in progress) crosses the boundary of one cell and moves into a neighboring cell

- to improve the global interference level
- to improve load balancing between adjacent cells

The main criterion for call handoff to avoid dropped calls is the quality of transmission for the ongoing connection, both uplink and downlink. Similarly, the criterion for handoff to optimize global interference is the uplink and downlink transmission quality corresponding to each neighboring cell to which the MS could potentially be handed over. Handoffs for traffic load balancing are determined by the BSC and MSC based on information on congestion in various cells under their control. Such handoffs represent top-down decisions and may involve a number of MSs and cells within the system.

The transmission performance measures that can be used for handoff decisions of the first two types include bit error rate (BER), the path loss over the radio channel, and C/I ratio. In GSM, the MS takes measurements of received transmission quality on channels within the serving cell as well as on channels in the neighboring cells, which are then reported (at least once every second) to the serving BSC for handoff decisions. The handoffs may be intra-BSC, inter-BSC, or inter-MSC, as shown in Figure 3.12. Call flows for an inter-MSC handoff are shown in Figure 3.13.

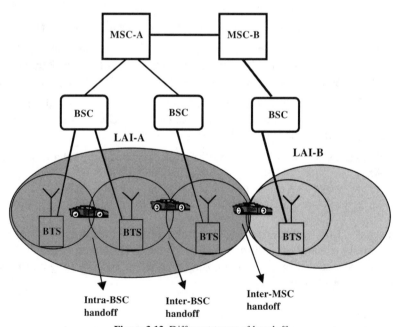

Figure 3.12 Different types of handoff.

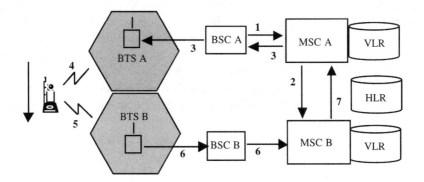

MS moves from cell A to cell B

1. BSC A informs MSC A that MS needs handover from BTS A to BTS B.

2. MSC A informs MSC B that a handover from BTS A to BTS B is underway.

3. MSC A commands BSC A/BTS A to proceed with handover to BTS B.

4. BTS A commands MS to change to a specified channel on BTS B.

5. MS informs BTS B that it is on specified channel on BTS B.

6. BTS B informs BSC A/MSC A that handover is complete.

7. MSC B informs MSC A that handover to BTS B is complete.

Note: MSC A continues to maintain control of call routing and connection

Figure 3.13 Call flow for inter-MSC call handoff.

3.2.7 Evolutionary Directions for GSM

Though the initial GSM concept anticipated that GSM would support a large set of services, phase 1, the self-contained standard on which the first commercial GSM service was launched, supports only a subset of services. GSM phase 2 specification, completed in 1995, represents the full-fledged version of GSM, which will support a number of additional supplementary services. It also specifies enhancements to the Mobile Application Protocol that will facilitate backward compatibility (with phase 1), and forward compatibility with GSM services and functions beyond phase 2. The list of services supported by GSM phases 1 and 2 are summarized in Section 3.2.1. During this period a half-rate codec was also specified to increase the spectral efficiency.

The phase 2+ activity under way in ETSI SMG is addressing a number of services and capabilities that can be added gradually and on an incremental basis as the market demand is identified. More than 80 items have so far been identified for phase 2+, and these represent evolution of GSM in such areas as speech coding, group call services, packet radio, high speed data, and IN and supplementary services. Some of the features addressed under phase 2+ in ETSI SMG are described in Sections 3.2.7.1 to 3.2.7.6.

3.2.7.1   New Supplementary Services.   A number of new supplementary services are being considered as part of phase 2+ activities. These services include the following:

- call completion to busy subscribers (CCBS)
- private numbering plan
- call deflection services
- mobile access hunting
- explicit call transfer
- multiple subscriber profile
- malicious call identification

3.2.7.2   New Features and Capabilities.   The following features/capabilities are also part of the phase 2+ work that is under way:

- optimum routing
- GSM/DCS1800 roaming
- support of universal personal telecommunication (UPT)
- satellite interworking
- access to GSM for digital enhanced cordless telecommunications (DECT)
- enhanced SIM capability (SIM tool kit)
- enhanced full-rate (EFR) speech coder

3.2.7.3   CAMEL (GSM/IN Integration).   The key thrust for the CAMEL (customized applications for mobile network enhanced logic) feature is to enable GSM to provide flexible service creation and export of operator-specific services to roaming subscribers. Though full usage of IN in GSM is identified as a long-term goal, the scope of CAMEL represents a first step to rapidly introduce short-term needs of the GSM network operators to provide IN services to their subscribers independent of their location. The proposed architecture for CAMEL phase 2 is illustrated in Figure 3.14. It is envisaged that the service switching function (SSF) will be integrated with the MSC functions in a combined MSC/SCP node. However, CAMEL phase 1 does not provide a specialized resource function for GSM (gsmSRF).

3.2.7.4   High Speed Circuit-Switched Data (HSCSD).   HSCSD, an enhancement to basic services provided by GSM, addresses the need for higher bit rate data services. Currently GSM data services are limited to user rates up to 9.6 kb/s. This is considered insufficient for emerging data services requirements. HSCSD is intended to provide data rates up to 64 kb/s by combining several 9.6 kb/s communications to provide one high rate communication. Chapter 5 describes in greater detail the range of data services provided and planned for GSM, including GPRS and HSCSD services.

3.2.7.5   General Packet Radio Service (GPRS).   GSM already offers the possibility to support packet data service for accessing public packet-switched networks via a dedicated traffic channel. It also offers short message service, which is true packet radio service (for short messages), and uses con-

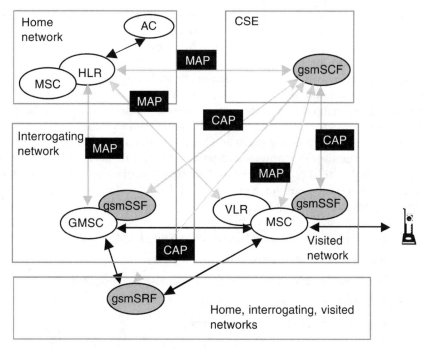

CSE	CAMEL service environment
CAP	CAMEL Application Protocol
MAP	Mobile Application Protocol

Figure 3.14 Proposed CAMEL phase 2 architecture.

trol and signaling channels on a shared basis. The aim of GPRS is to fulfill the need for a packet service that can provide efficient access to IP and X.25 services with data rates up to 115 kb/s.

3.2.7.6   Migration to a Universal Mobile Telecommunication System (UMTS).   UMTS is the next (third) generation mobile system that is being developed by ETSI for possible commercial deployment around year 2005. There is an emerging consensus that GSM network should migrate/evolve so that it can provide the defining functions/capabilities of UMTS. ETSI SMG is also responsible for the development of UMTS specifications.

3.3   IS-136: THE NORTH AMERICAN TDMA DIGITAL CELLULAR STANDARD (D-AMPS)

3.3.1   Background on North American Digital Cellular

The key objectives for the GSM standard in Europe were to specify a single, high capacity digital cellular system for Europe to replace the existing, incompatible analog systems that inhibited roaming across Europe. The GSM was also assigned new spectrum in the 900 MHz band for this purpose. North America has had the benefits of an analog cellular system based on a single (AMPS) standard with wide-area roaming provided by the IS-41 standard for intersystem operation. However, it was foreseen that the existing analog system would be unable to meet the rapidly increasing demand for cellular telephone services—especially in high density metropolitan areas.

The development of the digital cellular systems in the United States was triggered by the user performance requirements (UPR) documents produced by the Cellular Telecommunications Industry Association (CTIA) in 1988, which expressed the need for a new all-digital air interface design that would achieve goals that included the following:

- tenfold increase in capacity over the analog (AMPS) system
- ease of transition and compatibility with the analog system
- short time to service and low cost for dual-mode operation
- enhanced security, privacy, and quality
- new and enhanced services/features

The primary aim of the digital cellular standards in North America was therefore to increase the capacity of the existing spectrum and to provide a wider range of services with improved performance. However, in the process of achieving the aims, the North American cellular industry must face the need to support four different and somewhat incompatible standards—two analog (AMPS and NAMPS) and two digital (TDMA and CDMA)—using dual-mode terminals with AMPS as a fallback protocol.

As mentioned above, two digital radio standards have been specified for North America by the Telecommunication Industries Association (TIA). The IS-136 standard is based on TDMA radio access technology, and the IS-95 standard is based on the CDMA radio access technology. These digital cellular standards require that the present analog system (AMPS) and the new digital systems coexist (using dual-mode terminals sharing a common frequency band), with one or more digital systems to completely replace the analog AMPS system by the end of the twentieth century. The structure of TR45, the TIA committee that is primarily involved in the development of North American cellular standards in the 800 MHz band, is summarized in Figure 3.15. TIA Committee TR46 addresses mobile and personal communication standards

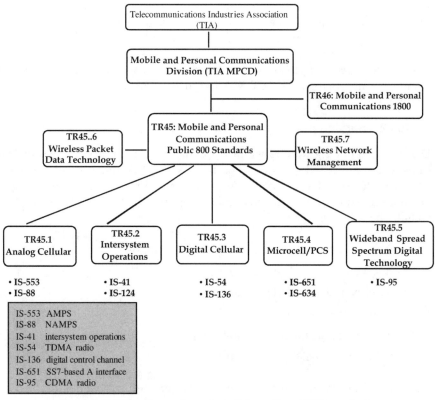

Figure 3.15 TIA subcommittees for cellular radio and PCS standards.

for the 1800 MHz band and has the responsibility for the North American GSM-based PCS1900 standard. This group works closely with Committee T1P1 (under Alliance for Telecommunications Industry Standards, or ATIS) and the ETSI SMG for PCS1900 related standards. Besides the work related to North American mobile and personal communication standards, both TIA TR45 and TIA TR46 develop North American inputs for IMT-2000, the international third-generation wireless standard being drafted in the Radio and Standardization Sectors of the International Telecommunications Union (ITU).

The IS-54-B standard for D-AMPS was completed in 1992, and commercial systems based on this standard are now in service in the United States, Canada, Mexico, Brazil, Kenya, Israel, and other countries. In late 1994 the capabilities of the IS-54 TDMA system were significantly enhanced when the IS-136 standard on a digital control channel (DCC) was completed. The availability of the DCC not only improves the control channel signaling capability of D-AMPS but also offers such enhanced features as point-to-point short messaging service, broadcast messaging service, group addressing, closed user groups, hierarchical cell structures, and slotted paging channel to support a *sleep mode* in the terminal (to conserve battery power). The IS-95 standard for digital CDMA systems was completed in 1993, and commercial services using the CDMA technology have been implemented in North America as well as many other countries and regions. The CDMA-based cellular system is described in Section 3.5.

3.3.2 Service Aspects for D-AMPS (IS-136)

D-AMPS supports a wide range of basic and supplementary services or features. These are listed and described in TIA IS-53-A and IS-53-B specifications. Besides such basic services as speech, asynchronous data, and group 3 fax, the service features and capabilities supported by the D-AMPS standard include the following:

- short message service (SMS)—point to point
- emergency (E.911) service
- lawfully authorized electronic surveillance (LAES)
- on-the-air activation (OTA)
- sleep mode terminal operation
- rolling mask message encryption

Further, the D-AMPS will offer a complete range of supplementary service categories, which include:

- Call forwarding (CF) services
    CF unconditional
    CF busy
    CF no answer
    CF default

- Call termination services
    call delivery
    call waiting
    calling number identification presentation
    calling number identification restriction
    do not disturb
    flexible alerting
    message waiting notification
    mobile access hunting

- Call origination services
    preferred language
    priority access and channel assignment (PACA)
    remote feature control
    voice mail retrieval

- Multiple-party services
    call transfer
    conference calling
    three-way calling

- Call restriction services
    password call acceptance
    selective call acceptance
    subscriber PIN access
    subscriber PIN intercept

- Privacy services
    voice privacy
    signaling message encryption

D-AMPS also provides asynchronous data service (ADS) by means of appropriate modems. Mobile originated service may be either "data only" or "voice then data." Mobile-terminated service may be of three types: "one number per service" (separate number for data)," two-stage dialing," or "voice then data."

3.3.3   Network Reference Model

Figure 3.16 presents the functional entities and the associated interface reference points for the North American digital cellular system. The network reference model is the same for both the TDMA-based (IS-136) and CDMA-

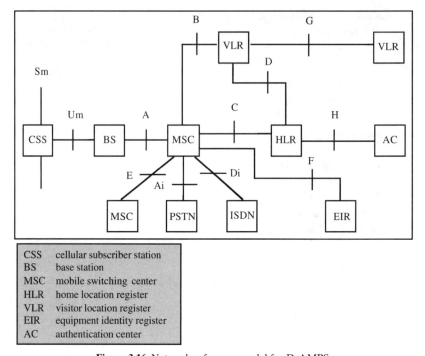

Figure 3.16 Network reference model for D-AMPS.

based (IS-95) radio systems. The model is used as a basis for specifying the messages and protocols for intersystem operation based on the TIA Interim Standard IS-41. The distribution of various functions within the physical network entities can vary among different implementations (e.g., the VLR functions are generally collocated with the MSC). If a number of functions are combined in a single physical equipment, the relevant interface reference points become internal and need not conform to the (open) specification.

The reference architecture and interface reference points in Figure 3.16 are almost identical to those for GSM (Figure 3.2). The functions assigned to each of the entities (BS, MSC, HLR, VLR, AC, EIR, etc.) essentially support the same functions supported in GSM, as described in Section 3.2.2. Note that in GSM, the base station is further partitioned into the base station controller (BSC) and base transceiver stations (BTS), with a defined (but rarely implemented) Abis interface.

Notwithstanding these similarities in the reference architecture and interface reference points, the underlying protocols and messages deployed in the D-AMPS for intersystem operation are quite different from those of the

GSM. For D-AMPS, interfaces, B, C, D, and E are fully defined in IS-41 (Revision C), and an SS7-based A interface (similar to the GSM A interface) has also been standardized (TIA IS-651).

3.3.4   Radio Aspects

D-AMPS will utilize the currently allocated spectrum for analog AMPS: that is, a total of 50 MHz in the 824–849 MHz (uplink) and 869–894 MHz (downlink), with each frequency channel assigned 30 kHz spacing. Each frequency channel then is time-multiplexed with a frame duration of 40 ms, which is partitioned into six time slots of 6.67 ms duration. The multiplexing and frame structure used in D-AMPS is shown in Figure 3.17, and the associated radio parameters are summarized in Table 3.3.

For mobile-to-base communication in a D-AMPS system using a full-rate codec, three mobiles transmit to a single base station radio by sharing the same frequency. To accomplish this, each mobile transmits periodic bursts of information to the base station in a predetermined order (mobile 1 on time

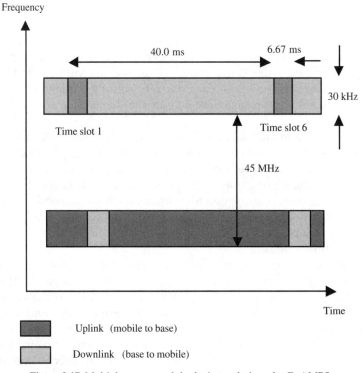

Figure 3.17 Multiple-access and duplexing technique for D-AMPS.

Table 3.3 Radio Parameters and Characteristics for D-AMPS

System Parameter	Value (IS-136)
Multiple access	TDMA/FDD
Uplink frequency (mobile-to-base)	824–849 MHz
Downlink frequency (base-to-mobile)	869–894 MHz
Channel bandwidth	30 kHz
Number of channels	832
Time slots/frame	6
Frame duration	40 ms
Interleaving duration	27 ms
Modulation	pi/4DQPSK
Speech coding method	VSELP convolutional
Speech coder bit rate	13.2 kb/s (full rate)
Associated control channel	Same frame
Handoff scheme	Mobile-assisted
Mobile station power levels	0.8, 1, 2, 3 W

slots 1 and 4, mobile 2 on time slots 2 and 5, mobile 3 on time slots 3 and 6. When the half-rate codec is implemented, six mobiles will be able to share a frequency channel by means of each one transmitting sequentially on the six available time slots, an arrangement that doubles the capacity of the D-AMPS system.

In the base-to-mobile communication link, each of the three mobiles (full-rate codec case) again shares the same (downlink) frequency channel. Each mobile receives the same information from the base station in the form of a continuous stream of time slots rather than periodic slots (bursts) of information. This continuous transmission is necessary for the dual-mode terminals to correctly synchronize to the TDMA transmission without the need for high accuracy synthesizers (clock) in the mobiles. Each mobile extracts, or reads, only its assigned time slot(s).

The speech coding and modulation process in the D-AMPS system is illustrated in Figure 3.18. The digitized speech at 64 kb/s is passed through a (vector sum excited linear predictive (VSELP) coder. The coder reduces the data rate to 7.95 kb/s, which passes through a channel coder and interleaver that increases the bit rate to 13 kb/s, which now includes error correction and detection bits, control channel data, training sequence data, and guard bits; the data are interleaved over two time slots (for full-rate speech coding). The channel coded and interleaved bit stream is then input to the pi/4DQPSK

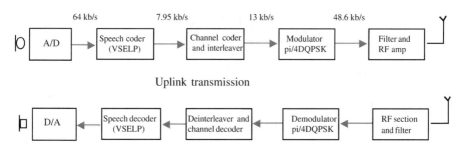

Uplink transmission

Downlink transmission

Figure 3.18 Speech coding and modulation in D-AMPS.

(Π/4- shifted **d**ifferential encoded **q**uadrature **p**hase **s**hift **k**eying) modulator, which modulates the data stream to a carrier in the 900 MHz band.

The slot format for the digital control channel (DCC) specified in TIA standard IS-136 is shown in Figure 3.19. Each slot is 6.67 ms in duration. The 16-bit PREAM field carries no information and is used by the base station to set the receiver amplifier to avoid signal distortion. The base station uses the bit pattern in the SYNC field to delineate the start of the incoming TDMA burst. SYNC+represents a fixed bit pattern, which provides additional synchronization information for the base station. The SCF field in the downlink slot provides information to support the random access scheme, and the CSFP helps the MS to find the start of the superframe.

Uplink

G	R	PREAM	SYNC	DATA	SYNC+	DATA
6	6	16	28	122	24	122

Downlink

SYNC	SCF	DATA	CSFP	DATA	SCF	RSVD
28	12	130	12	130	10	2

G	guard time	SYNC+	additional synchronization bits
R	ramp time for MS transmitter	SCF	shared channel feedback
PREAM	preamble	CSFP	coded superframe phase
SYNC	synchronization bits	RSVD	reserved
DATA	layer 3 payload		

Figure 3.19 Slot format for digital control channel.

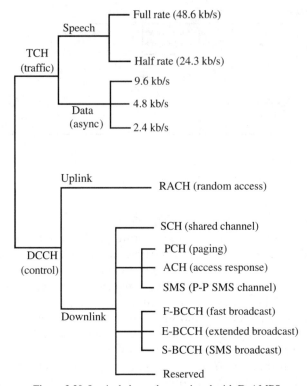

Figure 3.20 Logical channels associated with D-AMPS.

The digital control channel provides a number of logical channels for various control operations from base station to mobile. Figure 3.20 shows the logical channels associated with D-AMPS when the DCC is used.

3.3.5 Security Aspects

As in the case of GSM, D-AMPS uses a challenge response procedure based on a private key for authentication and privacy. However, the authentication and privacy algorithm, the nature of the private key, and the procedures for generating and transporting the authentication results for verification are different from those used in the GSM. A general description of the procedure specified in IS-41 is summarized here.

- At the time of subscription the MS is programmed with information specific to the subscriber or the terminal, such as mobile identification number (MIN) and electronic serial number (ESN), as well as the cellular authentication and voice encryption (CAVE) algorithm. Since the

D-AMPS currently does not utilize a subscriber identity module (as in the case of GSM), the private key (called the A key) is provided to the subscriber through a secure means (e.g., through registered mail). The subscriber then uses the terminal's keypad to enter the 64-bit A key into the MS, and its correct entry is verified by the security software within the MS. The A key also resides in the HLR/AC in the subscriber's home network.

- Once the subscriber-specific data, the CAVE algorithm, and the A key have been successfully programmed into the MS, the HLR/AC asks the MS to generate the secret shared data (SSD) by sending a RANDSSD (random number for SSD generation) parameter to the MS. This may take place when the MS makes the initial registration request. The MS utilizes RANDSSD, the A key, and the ESN as input to the CAVE algorithm to generate the SSD. This SSD is then used for generating authentication results and cryptographic keys. The SSD is also resident at the HLR/AC in the home network. The SSD for the specific MS may be changed at the discretion of the service provider (e.g., at fixed intervals or when fraudulent use of the terminal is suspected).

- To allow the visited network to authenticate a roaming subscriber autonomously, the SSD is passed from the HLR/AC of the home network to the serving VLR, along with the subscriber-specific data (MIN/ESN). This transfer takes place when the subscriber first roams into the new network and invokes reregistration/location update procedures. If the roaming agreement (between the home and the visited networks) does not include sharing of SSD, authentication response may be verified and crypto keys generated at the HLR/AC and sent to the VLR. However, it is more efficient if SSD is shared.

- In current implementations, the VLR broadcasts a *global challenge* at frequent intervals in the form of a 32-bit random number (RAND), which can be used by any terminal served by the VLR to generate necessary authentication result. The use of the global challenge (as opposed to a *unique challenge* to individual terminals) enables the MS to respond to the challenge and send the result as part of a service request (e.g., call setup), thereby eliminating additional messaging on the radio channel. The authentication result (an 18-bit response called AUTHR) and the ciphering key at the MS are generated by activating the CAVE algorithm, using RAND (global challenge), SSD and MIN/ESN.

- When the call setup request message (containing the AUTHR) from the MS is received by the MSC/VLR, the VLR also generates its own AUTHR using RAND, SSD, and MIN/ESN for the MS and compares it with AUTHR from the MS. If the two match, the MS is authenticated and the *start ciphering* command (with transfer of ciphering key to the base station) is issued.

- An unauthorized interception of the SSD during its transport from the HLR to the VLR may result in the impersonation of a user and fraudulent use of the network. In D-AMPS a *call count* is used to prevent such fraud as well as general MS duplication or *cloning*. The call count is incremented in the MS upon a request from the network—generally during a call. The network also maintains the count. If multiple mobile stations are sharing an identity, the network will notice the discrepancy and take necessary action (e.g., changing the SSD).

- During network access and other times (at the discretion of the network), a unique challenge may be used to verify the authenticity of the terminal/user. However, the unique challenge/response procedure does not generate cryptographic keys. Authentication for registration and call termination use ESN and MIN, whereas call originations use ESN

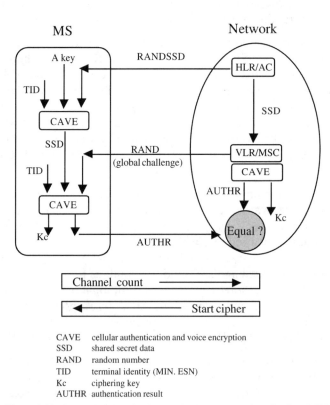

Figure 3.21 General authentication and privacy procedure for D-AMPS.

and a subset of the dialed digits, the latter replacing the MIN as input to the CAVE algorithm.

The general authentication and privacy procedure used in D-AMPS is illustrated in Figure 3.21, which assumes shared SSD. Confidentiality for the users in terms of their identity and location is an important security aspect in mobile networks, where the user identity may be transported over an unencrypted radio channel during initial registration or call setup. In the GSM, this threat is addressed by assigning a temporary, local identity (TMSI) for use across the radio channel. No such mechanism is currently available in D-AMPS, though with the recent adoption of IMSI for D-AMPS (IS-41,-Revision D), this potential security threat will be minimized.

3.3.6   Protocol Model and Typical Call Flow Sequences

Whereas the GSM specifications provide complete definitions of the Abis interface (between BTS and BSC) and the A interface (between the BSC and the MSC), till recently these interfaces were not defined for D-AMPS. The A interface based on the SS7 signaling transport has now been specified, but currently there are no plans to specify an interface equivalent to Abis in GSM. The simplified protocol reference architecture for D-AMPS is shown in Figure 3.22.

Functionally the protocol model for the A interface and the network interfaces (e.g., between the MSC and the HLR) are very similar in that they uti-

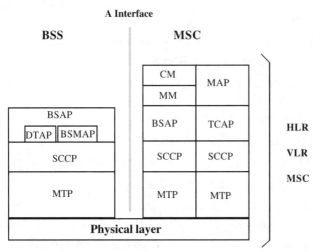

Figure 3.22 Protocol reference model: D-AMPS.

lize the same SS7 layers for signaling transport. In the United States, however, such SS7 entities as the MTP, SCCP, and TCAP, used in D-AMPS specifications, are based on the ANSI SS7 standards (i.e., ANSI standards for MTP, SCCP, and TCAP) as opposed to the ITU-T SS7 standards (used in the GSM system). Further, the Mobile Application Protocol (MAP) used in D-AMPS

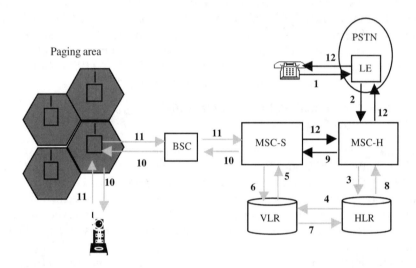

1. The PSTN user dials the phone number of the called mobile.
2. The serving CO routes the call to the Home MSC of the called mobile.
3. MSC-H sends an IS-41 Location Request to the HLR using called mobile's MIN.
4. The HLR sends a Route Request message to the VLR serving the mobile.
5. The VLR passes the Route Request message to the serving MSC.
6. MSC-S allocates a TLDN (Temporary Local Directory Number) and sends it to the VLR.
7. The VLR sends the TLDN to the HLR.
8. The HLR sends the TLDN to MSC-H.
9. MSC-H uses the TLDN to route the call to MSC-S.
10. MSC-S pages the mobile (over the paging area).
11. Called mobile responds to the page.
12. Calling and called parties are connected and the call proceeds.

Figure 3.23 Call flow for call delivery to a mobile subscriber in D-AMPS: dotted lines, signaling; solid lines, traffic.

was developed by the TIA as part of IS-41 standard to facilitate cellular system interoperability and roaming. Currently IS-41 also allows the use of X.25-based signaling (data) transport (i.e., X.25 physical layer, data link layer, and network layer), which then supports the IS-41 MAP.

The call sequences for D-AMPS are similar to those presented for the GSM system in Section 3.2.6. However, unlike the GSM system, D-AMPS currently does not use a gateway MSC or the IMSI. In place of IMSI, the North American cellular systems currently use the mobile identification number (MIN), which is the binary-coded version of the mobile subscriber's phone number (see Appendix A on Numbering and Identities). Enhancements in the D-AMPS standards have been completed to support IMSI. The messages and the parameters used to populate them are those defined in the IS-41 standard. Figure 3.23 provides an example call flow for call delivery to a mobile subscriber in the D-AMPS system.

3.3.7 Evolutionary Directions

The base standard defining the D-AMPS system is IS-54, along with IS-136, which enhances IS-54 by providing additional capabilities through the use of the digital control channel, and IS-41, which supports intersystem operations for wide-area roaming. As in the case of GSM, these standards are being enhanced on a regular basis by TIA TR-45 to support a wider range of services and capabilities. The set of services to be implemented in D-AMPS is described in IS-53 and IS-53B, which in turn were summarized in Section 3.3.2. Additional services/features that are being specified for D-AMPS include the following:

- packet- and circuit-switched data services (9.6, 14.4, and 64 kb/s) to support such applications as Internet and intranet access, group 3 fax, point-of-sale (POS) transactions, and telemetry
- short message services
- smart card
- emergency (E.911) service
- over-the-air service provisioning (OTASP)
- enhanced variable-rate codec (EVRC)

Further, for cellular and PCS providers to create and provision a wide range of services quickly and efficiently, and to offer the services to their subscribers in a roaming environment, IN capabilities are being added as part of the wireless IN (WIN) standardization activities under TIA TR-45.2. The WIN standard complements and interworks with IS-41 standard, and therefore the WIN capability will be available in not only D-AMPS but also in all other systems that utilize the IS-41 networking standard (e.g., AMPS and IS-95 CDMA).

The WIN standard may therefore be considered to be the North American equivalent of the CAMEL specification for GSM. The current WIN standard is focused on adding the necessary trigger points and enhancing the call model (in line with the ITU-T IN standards), and thus it was possible to introduce IN functions and concepts in North American cellular systems while maintaining maxi-

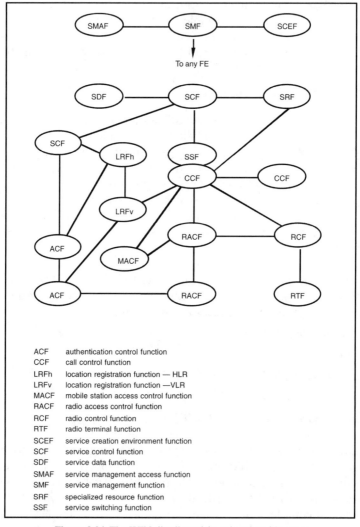

Figure 3.24 The WIN distributed functional architecture.

mum backward compatibility (e.g., use of IS-41 Mobile Application Protocol). The WIN functional architecture adopted by the TIA TR45.2 is shown in Figure 3.24. For actual implementation of WIN services, the functional entities (FEs) shown in Figure 3.24 will need to be mapped onto appropriate network (physical) entities associated with the ANSI-41 network and the intelligent network. A simplified example of such a physical architecture is shown Figure 3.25.

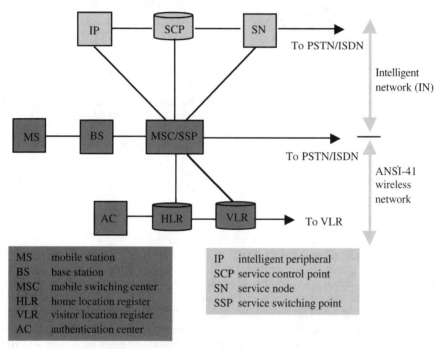

Figure 3.25 An example of a physical architecture for WIN.

The longer term enhancements to IS-136 D-AMPS are aimed at moving its capabilities toward the next (third) generation wireless systems as defined by the emerging IMT-2000 standard in the ITU. The evolution from current IS-136 capabilities (phase 1) to third generation is planned in two steps. The objectives of phase 2, also known as IS-136+ include:

- higher bearer rates (enhanced modulation techniques)
- higher spectral efficiency
- improved voice quality (toll quality)
- higher speed (up to 64 kb/s) circuit- and packet-switched data (similar to HSCSD and GPRS in GSM)

The primary objective of phase 3 for D-AMPS evolution, also known as IS-136HS (high speed) is to support third-generation data capabilities, which include 144 kb/s for vehicular, 384 kb/s for pedestrian, and 2 Mb/s for the indoor environments. Whereas IS-136+ will continue to use 30 kHz carriers, IS-136HS will utilize 200 kHz carriers for outdoor and 1.6 MHz carriers for indoor environments to support targeted higher data rates. Since GSM and D-AMPS utilize similar TDMA radio transmission technologies, some level of cooperation and collaboration in their evolution efforts is anticipated. For example, concepts like GPRS and EDGE (general radio packet service, enhanced data rates for GSM evolution), developed for GSM, are likely to be adopted for IS-136 D-AMPS.

TIA TR45 has submitted the radio transmission technology (RTT) proposal based on the evolution of IS-136 to the ITU-R for consideration as a potential radio interface for IMT-2000, the international third-generation wireless standard. This RTT proposal is identified as UWC-136, where UWC refers to the Universal Wireless Communications Consortium, consisting of AMPS and D-AMPS operators and equipment vendors from around the world.

3.4 PDC: THE JAPANESE TDMA DIGITAL CELLULAR STANDARD

3.4.1 Radio Aspects of PDC

In April 1991 the Research and Development Center for Radio Systems of Japan completed the detailed standard for the Japanese digital cellular system currently known as PDC (personal digital cellular). The standard specifies a unified air interface for the digital cellular system in Japan. The air interface is based on TDMA, and as shown in Table 3.4, it has significant similarities to the D-AMPS system in North America.

The PDC radio interface has certain key features which the system designers believe make it superior to the GSM system. These features include the following:

- PDC employs diversity reception in the mobile stations which obviates the need for equalizers, which are an essential component of GSM.

- PDC uses a much lower transmission bit rate (42 kb/s vs. 270.83 kb/s in GSM), which leads to better spectrum utilization, higher capacity, and lower cost.

- The access signaling protocols in the PDC are simpler and require fewer procedures.

Table 3.4 Radio Characteristics for PDC, D-AMPS, and GSM

Parameter	PDC	D-AMPS	GSM
Forward channel	810–826 MHz	869–894 MHz	935–960 MHz
	1477–1489 MHz		
	1501–1513 MHz		
Reverse channel	940–956 MHz	824–849 MHz	890–915 MHz
	1429–1441 MHz		
	1453–1465 MHz		
Carrier spacing	25 kHz interleaving	30 kHz interleaving	200 kHz interleaving
Modulation	pi/4 DQPSK	pi/4 DQPSK	GMSK
Channel bit rate	42 kb/s	48.6 kb/s	270.8 kb/s
Codec (full rate)	11.2 kb/s (VSELP)	13 kb/s (VSELP)	13 kb/s (RPE-LTP)
Frame duration	20 ms	20 ms	4.6 ms
Channels/carrier	3 (full rate)	3 (full rate)	8 (full rate)
	6 (half rate)	6 (half rate)	16 (half rate)
MS power levels	0.3, 0.8, 2 W	0.8, 1, 2, 3 W	0.8, 2, 5, 8 W

The channel structure used in PDC is shown in Figure 3.26. Control channels in the PDC consist of common access control (CAC) and user-specific control (USC) channels. The common access channel consists of a broadcast channel (BCCH), a common control channel (CCCH), and a user packet channel (UPCH). The common control channel is further classified into a paging and a signaling channel. Base stations use individual frequencies for common access (CAC) with paging and signaling channels mapped to the same physical channel. UPCH is used for packet services. SACCH (slow channel) is used for radio link control during the call, and FACCH (fast) carries control information by briefly interrupting the call in progress.

3.4.2   Signaling Structure in PDC

The signaling structure for the PDC was developed to ensure efficient spectrum utilization, support of enhanced services, and alignment with the open system interconnection (OSI) model, and the ITU signaling recommendations. As shown in Figure 3.27, the signaling structure is divided into three layers. Layer 1 (physical layer) addresses such functions as collision control, error control, cyclic redundancy check (CRC), and segmentation/reassembly. Layer 2 consists of an address part and a control part; the former is used to allow a

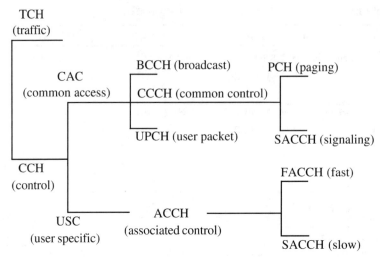

Figure 3.26 Channel structure for the PDC system.

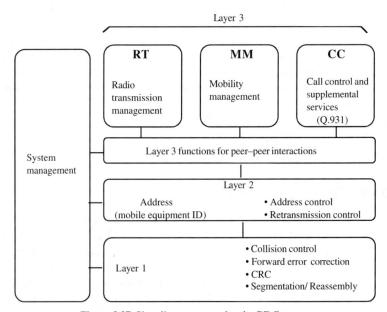

Figure 3.27 Signaling structure for the PDC system.

base station and multiple mobile stations to exchange link control information using the common control channel. The control part is used for retransmission of corrupted signals.

Layer 3 is partitioned into radio transmission management (RT) or radio resource control, mobility management (MM), and call control (CC). The call control procedures for basic call setup and takedown and supplementary services are based on the ITU-T ISDN signaling protocols (Q. 931). However, to avoid increased delays due to low speed of the associated control channel (1 kb/s), a shortened procedure specific to mobile radio is provided in addition to the Q.931 procedures.

3.4.3   PDC Network Configuration

The PDC standard essentially specifies the radio interface and leaves the individual operators to decide the network configuration they wish to implement. The most commonly deployed configuration for PDC network (by NTT DoCoMo) is shown in Figure 3.28. The network configuration

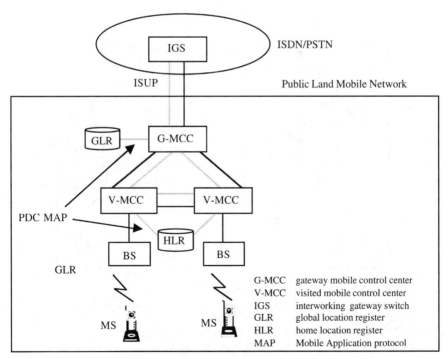

Figure 3.28 Network configuration for the PDC system.

provides connectivity to fixed networks (PSTN, ISDN, PSPDN), ensures roaming between different PDC cellular networks, and uses a unified interface for interconnection of PDC cellular network and the fixed network.

The gateway mobile control center (G-MCC) is a toll switch that provides gateway functions to the fixed network(s). The visited MCC (V-MCC) provides call and bearer connection functions, and interfaces to an HLR containing up-to-date information on mobile station identities and locations. The location tracking of roaming mobile subscribers is assisted by the gateway location register (GLR), which temporarily stores the data related to mobile stations roaming into its domain. Steps involved in the delivery of a call to a roaming subscriber are shown in Figure 3.29.

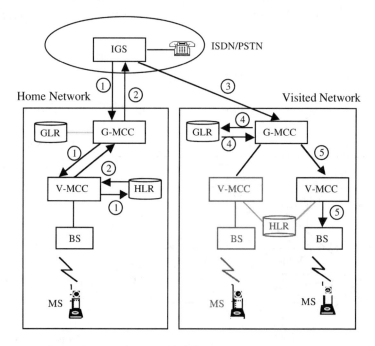

1   Request for current location of the MS.
2   Home network returns information on current location (serving G-MCC).
3   IGS forwards call to serving G-MCC.
4   G-MCC obtains information on V-MCC/BS serving the MS.
5   G-MCC issues paging request to serving V-MCC/BS.

Figure 3.29 Incoming call delivery to a mobile subscriber in PDC.

3.5   IS-95: THE NORTH AMERICAN CDMA DIGITAL CELLULAR STANDARD

3.5.1   Introduction

One of the targets set by the CTIA for the digital cellular systems for North America was a tenfold increase in traffic capacity over the analog AMPS system. The capacity increase provided by the TDMA-based D-AMPS, which currently supports three traffic channels per carrier, is only three times the AMPS capacity. In 1990 Qualcomm developed and demonstrated a CDMA-based digital cellular system that claimed a twentyfold increase in capacity over the analog system. In 1992, following a request from the CTIA, TIA TR45 initiated standardization work on wideband spread spectrum technologies for cellular applications. The IS-95 standard for the CDMA common air interface was adopted in 1993, followed by an enhanced and revised version (IS-95A) in 1995.

Commercial CDMA systems are now being implemented in North America and are also operational in many other countries. The final assessment on the potential superiority of CDMA systems over TDMA systems, in terms of capacity, cost, and speech quality, will emerge only after both systems have been in operation in dense, urban areas with full complements of subscribers and services.

3.5.2   Service Aspects

Some of the features and services that have been standardized as part of IS-95A standard include the following:

- *Short message service (SMS)*: will support both mobile-terminated and mobile-originated short messages of up to 255 octets on either the control channel or traffic channel. SMS can also be used to support such applications as telemetry and digital paging.

- *Slotted paging*: enables mobiles to wake up for one to two time slots (80 ms) per slotted paging cycle (1.28–1.64 s) to listen to incoming pages. The feature is intended to conserve mobile battery power.

- *Over-the-air activation (OTA)*: allows the mobile to be activated by the service provider without third-party intervention. It also provides potential for future remote reprogramming of mobile terminals and for software download capabilities.

- *Enhanced mobile station identities*: provides for use of international mobile station identities (IMSI) based on the ITU-T standard (E.212), thereby facilitating international roaming and separation of mobile DNs and mobile terminal identities.

- *Temporary mobile station identities (TMSI)*: allows for the allocation of TMSI by serving VLRs, which are used on the air interface to maintain user confidentiality and to reduce overall signaling traffic load across the radio interface.

- *Asynchronous data and group 3 fax*

- *Synchronous data*: applications for secure telephone unit (STU III).

- *Packet data*

- *Supplementary services*: call waiting, call forwarding, calling line ID., and similar services that are currently supported by D-AMPS and GSM.

3.5.3   Network Reference Model and Security Aspects

The IS-41 standard for internetwork operation has also been enhanced to support the network and signaling requirements for CDMA cellular systems. IS-95 CDMA systems also use the authentication and privacy procedures specified in IS-41 which are used in the D-AMPS systems. Thus the network reference model shown in Figure 3.16, and the security aspects addressed in Section 3.3.5 generally also apply for CDMA systems based on the IS-95 standard.

3.5.4.   Radio Aspects

The spread spectrum techniques used in CDMA cellular systems are adaptations of similar techniques, used extensively in military applications since 1950. In these techniques, the set of normal payload data is modulated and transmitted using a special spreading code. At the receiving end, the desired signal is recovered by despreading the signal with an exact copy of the spreading code in the receiver correlator. Other signals (within the same frequency band) remain fully spread and are perceived as noise.

The TDMA-based digital cellular systems like GSM and D-AMPS are examples of band-limited systems that aim to maximize the transmitted information rate within the allocated bandwidth by increasing the ratio of bit energy to noise power spectral density ($E_b N_o$) or the signal-to-noise ratio (SNR). An increase in bandwidth efficiency can be achieved either by selecting a modulation technique like DQPSK (used in D-AMPS and the Japanese PDC system) that is very agile (i.e., carries an increased number of bits per symbol) or by using a bandwidth-conserving modulation technique like GMSK (used in GSM).

A system based on the spread spectrum concept is an example of a power-limited system that is not constrained by bandwidth. In fact, in such systems the transmitted signal is much wider (in frequency spread) than that required for information to be carried. Though there are several kinds of spreading technique, such as direct sequence, frequency hopping, and chirping, the IS-95 CDMA cellular system employs direct sequence spread spectrum method. In these systems, the spreading is accomplished by modulating the narrowband information with much wider spreading signal provided by a pseudo–random noise code. Since each radio is assigned its own specific PN code, all radios except the desired one appear as noise at the receiver when the composite received signal is correlated with the PN code of the desired radio. The PN spreading code is often referred as the chipping code and the resulting bandwidth (after spreading) as the chip rate. In the IS-95 CDMA cellular systems the chip rate is about 1.23 MHz; approximately one-tenth of the total 12.5 MHz spectrum (for each direction) allocated to each cellular operator in the United States.

Thus, a set of ten 1.25 MHz bandwidth CDMA channels can be used by each operator if the entire allocation is converted to CDMA. However, in the dual-mode CDMA/AMPS environment that is likely to exist for the foreseeable future, initially only one or a few 1.25 MHz channels need to be assigned to CDMA digital service from the present AMPS analog service. Gradually as the demand for digital service increases, more and more analog capacity (in increments of 1.25 MHz) can be transferred to CDMA digital operation. In a mixed CDMA/AMPS environment, some frequency guard band is required (between the CDMA and AMPS frequency allocations) to ensure that maximum CDMA call-carrying capacity can be realized.

3.5.4.1   Forward Link Structure in IS-95.   The CDMA common air interface (CAI) specifies a forward physical channel (known as forward waveform) design that uses a combination of frequency division, pseudo–random code division, and orthogonal signal multiple-access techniques. As mentioned earlier, frequency division is achieved by dividing the available cellular spectrum into nominal 1.23 MHz channels by combining 41 AMPS channels of 30 kHz each. These 1.23 MHz channels can be increased from the initial single CDMA channel to multiple such channels as demand for digital service increases. An example of a logical forward (base station to the mobile) waveform for an IS-95 CDMA system is shown in Figure 3.30. It consists of a maximum of 64 channels, which include a pilot channel, a synchronization channel (optional), paging channels (maximum of 7), and traffic channels.

PILOT CHANNEL.   The pilot channel consists of an unmodulated direct sequence spread spectrum signal with its own identifying spreading code and is shared among all users (mobiles) in a sector or cell. Its data content is a sequence of zeros that are not channel-encoded. Each pilot transmits the same

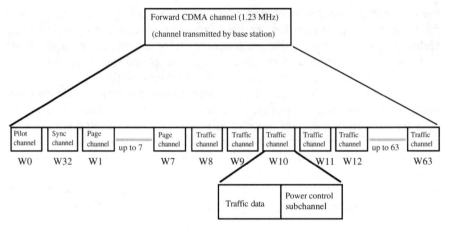

Figure 3.30 Structure of an IS-95 CDMA forward waveform.

spreading sequence at different time offset, which can be used to distinguish signals of different pilots. The basic uses of the pilot channel include the following:

- uniquely identifying sectors/cells
- providing phase/time/signal strength reference
- identifying multipath components
- identifying handoff candidates
- rescanning or periodically checking for a better sector/cell

Mobiles continuously monitor pilot channels while on a paging or traffic channel so that they can move to a different cell (with better signal).

SYNCHRONIZATION CHANNEL. The synchronization channel provides a mobile with the system parameters required to synchronize with the network and obtain a paging channel. The synchronization channel therefore includes such information as the following:

- system identification number (SID)
- network identification number (NID)
- version of the radio interface being supported
- precise time-of-day information
- PN offset of the associated pilot channel

The synchronization channel is optional in the sense that for small cells it may be omitted and required synchronization may be provided from a neighboring cell. The synchronization channel, which is created at 1200 b/s, is half-rate channel-encoded and repeated once to yield a 4800 b/s data rate, which is input to a bit interleaver.

PAGING CHANNELS.   A forward waveform transmitted by a base station can have up to seven paging channels, which contain messages to a specific mobile, as well as necessary system parameters. Mobiles are assigned paging channels by a hash function based on the mobile identification number (MIN). Typical messages on the paging channel include pages, traffic channel assignments, and short messages.

FORWARD TRAFFIC CHANNELS.   The forward traffic channels are used to transmit voice or data to a mobile that is in a call. There may be up to 63 forward traffic channels, depending on the number of paging channels and the presence of synchronization channels. Data rates are flexible (1200, 4800, or 9600 b/s) to support variable-rate voice coders, and these are structured in 20 ms frames. Signaling information from the base station to the mobile during a call can be transmitted using *blank-and-burst* or *dim-and-burst* methods.

The modulation process for the forward CDMA traffic channel is shown in Figure 3.31. The physical layer structure for the synchronization and paging channels is very similar to that of the traffic channels. Some of the key features include the following:

- Channels in the forward link are logically separated by the unique Walsh codes.

- Voice information is digitized by means of a quadrature code excited linear predictive (QCELP) variable-rate codec.

- Digitized information is uniformly protected by the rate 1/2 convolution code followed by block interleaving to achieve time diversity.

- Power control bit is injected every 1.25 ms in the forward traffic channels.

- Each base station is assigned a unique PN spreading sequence.

3.5.4.2   Reverse Link Structure in IS-95.   The reverse CDMA channel is used by all the mobiles in a cell coverage area to transmit to the base station. The reverse channel as it appears at the base station receiver (Figure 3.32), is a composite of all the outputs from all the mobiles in the base station's coverage area. At any instant in time, there may be m mobiles engaged in a call (carrying user traffic) and n mobiles trying to gain access to the system. Thus the reverse channel structure allows up to 62 different traffic channels and 32

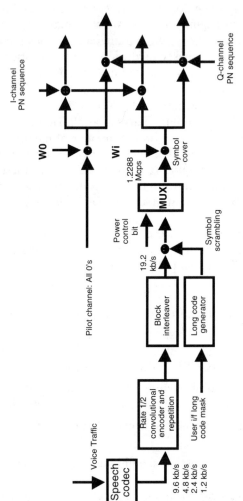

Figure 3.31 Modulation process for IS-95 CDMA forward traffic channels.

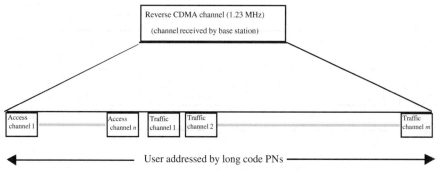

Figure 3.32 Structure of IS-95 CDMA reverse traffic channel.

different access channels. However, the number of channels in use at any time may be considerably lower.

ACCESS CHANNELS.   An access channel allows a mobile to communicate with the system when it needs to initiate an action (e.g., registration, call origination) or when it needs to respond to messages received on a paging channel. Since multiple mobiles may attempt access at the same time, the access channel utilizes some form of slotted random access protocol for contention management. The data rate for the access channel is 4800 b/s, and the number of access channels in a cell is configurable—generally one or two access channels for each paging channel. The access channels are spread by means of a long code determined by several parameters forming an identity unique to the access channel.

REVERSE TRAFFIC CHANNELS.   Reverse traffic channels are used for transmitting voice or data traffic from a mobile to the base station. Each of these channels is paired with a corresponding forward traffic channel. Signaling information during a call is again transmitted in *blank-and-burst* or *dim-and-burst* mode.

The modulation process for the reverse CDMA traffic channel is shown in Figure 3.33. The access channel operates at 4800 b/s. Some key features for the CDMA reverse link channels are as follows:

- Voice information is digitized using a QCELP variable-rate codec.

- All information bits are protected using rate 1/3 convolution code and block interleaving to achieve implicit time diversity.

- A stronger rate 1/3 convolution code is used to compensate for the absence of a pilot channel (which generally assists in coherent detection).

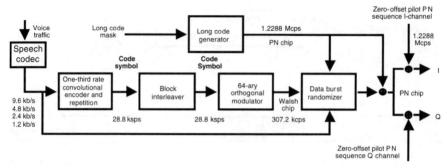

Figure 3.33 Modulation process for IS-95 CDMA reverse traffic channels.

- The reverse channels are tightly power-controlled by the base station to minimize interference.
- Each mobile station is assigned a unique PN spreading sequence.

Figure 3.34 summarizes the various channel types used in a CDMA cellular system and their characteristics.

Channel Type	Application	Quantity	Maximum Rate (b/s)	Spreading Code
Forward				
Pilot	System monitoring	1	NA	Walsh code 0
Synchronization	Synchronization	0 or 1	1200	Walsh code 32
Paging	Signaling (BS-to-idle MS)	≤ 7	9600	Walsh codes 1–7
Traffic	Voice/data (BS-to-MS)	≤ 63	9600/14,400	Walsh codes 8–31 and 33–63
Reverse				
Access	Signaling (idle MS-to-BS)	≤ 14*	4800	Access channel long code mask
Traffic	Voice/data (MS-to-BS)	≤ 63	9600/14,400	Mobile-specific long code mask

* Generally 1 to 2 access channels per paging channel.

Figure 3.34 Channel types in IS-95 CDMA cellular systems.

3.5.5   Some Key Features of IS-95 CDMA Systems

3.5.5.1   Diversity.   Cellular systems are prone to multipath fading and diversity methods of some form are required to mitigate the effects of fading. The types of diversity that are available in a CDMA system include the following:

- time diversity provided by symbol interleaving, error detection, and correction coding
- frequency diversity, provided by the 1.25 MHz wideband signal
- space (path) diversity, provided by dual cell-site receive antennas, multipath rake receivers, and multiple cell sites (soft handoff)

Whereas time and frequency diversity can be provided to a certain extent in FDMA and TDMA systems, a unique feature of CDMA systems is their ability to provide extensive path diversity and potential for improved performance in a difficult propagation environment. The path diversity is achieved through multipath processing by means of parallel correlators for the PN waveforms at the receivers in the mobile and the cell sites. Receivers using parallel correlators (rake receivers) allow signals arriving on individual paths to be tracked independently, and the sum of their received signal strengths is then used to demodulate the signal. The deployment of multiple correlators for the simultaneous tracking of signals from different cells provides the underlying basis for *soft handoff* in CDMA systems.

3.5.5.2   Power Control.   In a CDMA system, the mobile-to-base station link is subject to the so-called *near–far problem*, which considers the anomaly whereby a mobile station close to the base station has a much lower path loss than mobiles that are far removed from the base. If all the mobile stations were to use the same transmit power, the mobile(s) close to the base would effectively jam the signals from the mobiles far away from the base. Thus, for the CDMA system to work effectively, RF power in the system needs to be controlled. These power control requirements have two key components:

1 First, and most important, all the transmissions from the mobiles must be received at the base station's receiver at approximately the same strength (within 1 dB), even under conditions of fast multipath fading.

2 To maximize the number of users sharing a cell, only the minimum RF power required for reliable communication should be allowed from the base station transmitter.

To meet the first requirement, CDMA systems deploy open-loop and closed-loop controls. The open-loop control is used to counter the effect of wide dynamic range of the path loss (between the farthest and the closest mobile to the base station), which is in the order of 80 dB. In open-loop control, the mobile station estimates the loss between itself and the base station and then uses it to make a coarse adjustment in its transmitted RF power. The mobile estimates the path loss by measuring the received signal level (from the base station). The mobile then combines this reading with the power control information sent by the base station during some initial signaling transactions. This information includes the transmit power from the base station and some parameters that can be used to adjust the open-loop power control for different-sized cells and different cell ERP and receiver sensitivities.

Whereas open-loop power control attempts to provide a coarse adjustment to RF transmit power from a mobile, the closed-loop control mode actually measures the SNR at the cell-site receiver and, based on a predetermined threshold (desired SNR), instructs the mobile to adjust its transmit power. In practice, the cell receiver forms an estimate of the received SNR of the mobile's signal every 1.25 ms. The SNR measurement is compared with the set point or threshold value of the SNR. If the received SNR is too high (low), a decrease (increase) power command is sent to the mobile. This power control command is sent every 1.25 ms, providing an 800 b/s power control rate. The mobile responds to the power control bits (1, decrease power, 0, increase power) by making small adjustments of 2.0 dB in its RF power output until the received SNR at the cell site is within acceptable limits (within 1.0 dB of the desired set point).

To meet the second power control requirement (i.e., to minimize the base station transmit power and maximize the number of mobiles that may share a cell), the base station continuously lowers its transmit power in small steps until the responding mobile signals the base station for more power. This process is performed individually for each of the CDMA channels on a forward CDMA waveform.

3.5.5.3   *Soft Handoff.*

Soft handoff in a CDMA system results from the system's capability to simultaneously deliver signals to a mobile through more than one cell. Thus the handoff sequence in a CDMA system involves transition from the donor cell to both the donor and the receiving cell and finally to the receiving cell (similar to the handoff of a baton in a relay race). This *make-before-break* procedure not only reduces the probability of a dropped call during handoff, but it also makes the handoff virtually undetectable by the user. In this regard, the FDMA (analog) and TDMA (digital) based systems utilize a *break-before-make* call switching procedure, with potential adverse impact on quality of service. The soft handoff process in the CDMA system is illustrated in Figure 3.35.

The handoff is *mobile-assisted* in that the mobile station with a call in progress continues to monitor the signal strength from neighboring cells, if the

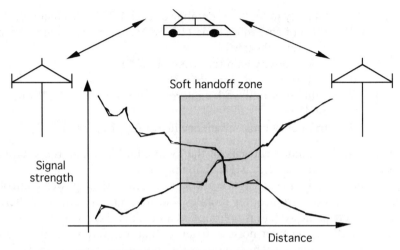

Figure 3.35 Soft handoff process in the CDMA system.

signal from one of these cells becomes comparable to that of the original cell, it initiates the handoff process by sending a control message to the MSC. The MSC responds by establishing a link to the mobile station through the new cell (identified in the control signal from the mobile), while maintaining the old link. While the mobile station is located in the transition region between two adjacent cells, the call can be supported by signals through both cells, thereby eliminating the common *border cell problems* of frequent back-and-forth hand-off (ping-ponging) between the two cells. The original cell will hand off the call only when the mobile station has been firmly established in the new cell.

   3.5.5.4 IS-95 CDMA System Capacity.   The key parameters that de-termine the capacity of a CDMA digital cellular system are as follows:

- processing gain (ratio of spreading code to information data rate (W/R)
- ratio of energy per bit to noise power (E_b/N_o)
- voice activity factor
- frequency reuse efficiency
- number of sectors in the cell-site antenna

   The theoretical capacity of the IS-95 CDMA cellular system in terms of calls per 1.25 MHz channel per cell is provided by:

$$N_p = \frac{(W/R)vS}{(E_b/N_o)F}$$

where N_p = capacity in terms of number of calls/1.25 MHz channel/cell

W/R = ratio of spreading code (1.2288 Mcps) to maximum information rate (9600 b/s)

v = voice activity gain (approximately 2)

S = sectors per cell (generally 3)

E_b/N_o = minimum ratio of bit energy to noise power (6 dB or factor
of 4)

F = frequency reuse efficiency factor (approximately 3/2)

The theoretical estimated capacity of the IS-95 CDMA system is therefore in the order of 128 calls per 1.25 MHz channel per cell.

The theoretical capacity figure above assumes perfect power control to counter interference from other mobiles and neglects the effect of thermal noise. In practice, the system is operated at a maximum capacity loading such that about half the receiver noise is from mutual interference and the other half is from thermal noise, which reduces the theoretical capacity above to approximately half (i.e., to 64 calls per 1.25 MHz channel per cell).

3.5.5.5 Soft Capacity.   CDMA radios based on the spread spectrum concept are designed to tolerate some level of interference, with their overall capacity limited by how well this mutual interference can be controlled. The capacity of conventional radios based on FDMA or TDMA concepts is limited by the number of noninterfering signals achieved by using complete separation either in frequency or time. Thus the capacity of CDMA systems is interference-limited, while the capacity of FDMA/TDMA systems is limited by the number of noninterfering coordinates in terms of frequency bands and/or time slots.

In the present U.S. cellular environment, each operator can deploy 12.5 MHz of spectrum, which provides 57 analog channels in a three-sector cell site. When demand for service is at a peak, the 58th request for a channel in the cell must be blocked and given the all-channels-busy signal. The CDMA systems, however, offer much softer relationship between the number of users and the (transmission) grade of service. In other words, an operator has the flexibility to admit additional users during peak periods by providing a somewhat degraded grade of service (increased bit error rates). This capability is especially important when calls might be dropped during handoff because of a lack of available free channels.

3.5.6   Evolutionary Directions

The emerging implementations of CDMA systems are generally based on IS-95A and IS-95B specifications. The short-term evolutionary goals beyond IS-95A and IS-95B capabilities are reflected in the scope of the IS-95C specification, scheduled for completion by the end of 1999. The major objectives for IS-95C include [24, 25]:

- improved capacity, coverage, and efficiency
- efficient medium rate data services (65–144 kb/s)
- longer battery standby times (several hundred hours)
- efficient mobile-assisted hard handoffs (CDMA-to-CDMA, CDMA-to-AMPS, CDMA-to-GSM)

IS-95C will meet these objectives with minimal hardware modifications (limited to digital signal processing), and with full transparency to services currently provided under IS-95A and IS-95B systems (complete backward compatibility).

The planned improvements in capacity and coverage and provision of efficient medium data rates will be achieved by implementing the following changes to the forward and reverse links in IS-95C:

- Forward link changes, including
    faster and more accurate power control
    replacing BPSK by QPSK (with 128 Walsh function channels and stronger codes)
    reducing soft-handoff overhead to 70%
- Reverse link changes, including
    reducing power control latency
    using BPSK (pilot-aided) modulation (for long-term coherent reception and continuous transmission)
    using rate 1/4 code instead of rate 1/2 and 1/3

Besides the work on IS-95C for achieving overall improved capacity, coverage, and efficiency for CDMA systems, there are efforts under way (by Qualcomm) to develop a proprietary specification in the short term, which will be optimized for high data rate (HDR) packet communication services. The expected characteristics of this specification, sometimes referred as 95-HDR, include the following:

- Dedicated optimized 1.25 MHz channel
- High speed peak data rates up to 1.8 Mb/s
- Decoupling of data services from voice services
- Maintaining high degree of compatibility with IS-95
    same chip rate (1.2288 Mcps)
    same power requirements as current terminals and base stations
    same link budgets
    same RF, IF, and baseband analog components
- Potential for integration with IS-95

The longer term objectives for IS-95 based CDMA systems are aimed toward meeting the requirements for IMT-2000, the third-generation wireless standard being specified by the ITU. However, these objectives are to be achieved by taking a graceful evolutionary path and maintaining full backward compatibility. Generally known as cdma2000, this IMT-2000-compliant CDMA system will reuse and build on the full complement of existing CDMA air interface and network standards. The key characteristics cdma2000 are as follows:

- channel sizes of 1.25 MHz (IX) and 3.75 MHz (3X)
- support for direct spread and multicarrier forward link
- physical layer support for all data rates in the 1X and 3X channels
- support for radio link protocol (RLP) for packet data
- backward compatibility with existing IS-95B data services (i.e., IS-707)
- backward compatibility with IS-95B voice services and signaling structures
- double the voice capacity of IS-95B
- extended terminal battery life through improved sleep mode
- ability to hand off between generations
- support for improved interfrequency hard handoff
- potential capability to operate with the IS-41- or GSM-based core networks

As described in Chapter 6, cdma2000 is one of the radio transmission technology (RTT) proposals being considered for IMT-2000 radio interface, and the defining radio parameters for cdma2000 are summarized in that chapter (see Table 6.2).

3.6   CONCLUDING REMARKS

The digital cellular systems based on TDMA and CDMA technologies are fully specified and are being implemented at a rapid rate around the world. The current activities in standards forums like ETSI (Europe), TIA (North America), and TTC (Japan) are directed toward enhancing the capabilities of these systems in terms of capacity, coverage, performance, and services and features. Use of low bit rate coders, importing IN concepts for better service delivery, and support of high bit rate circuit- and packet-switched services are a few of the areas in which significant progress has been made. The recognized need for a third-generation system that will provide global roaming and service delivery as well as support of high speed data, Internet, and multimedia services is playing a significant role in the evolution of these digital cellular systems, especially in terms of evolution toward the third-generation network and service requirements.

3.7 REFERENCES

[01] A. Mehrotra, *Cellular Radio-Analog and Digital Systems,* Artech House, Boston, 1994.

[02] M. Appleby and F. Harrison, "Cellular Radio Systems," Chapter 47, in *Telecommunication Engineer's Handbook,* F. Mazda, ed., Butterworth, Heinmann, London, 1995.

[03] M. Mouly and M.-B. Pautet, *The GSM System for Mobile Communications,* published by M. Mouly et Marie-B. Pautet, Palaiseu, France, 1992.

[04] M. Rahnema, "Overview of the GSM System and Protocol Architecture," *Communications Magazine,* Vol. 30, No. 12, December 1992.

[05] A. R. Potter, "Implementation of PCNs Using DCS1800," *Communications Magazine,* Vol. 30, No. 12, December 1992.

[06] M. Mouly and M.-B. Pautet, "Current Evolution of GSM Systems," *Personal Communications Magazine,* Vol. 2, No. 5, October 1995.

[07] P. Simmons and M. Mouly, "Switching Handovers in Microcellular Mobile Networks: An Architectural Evolution," International Switching Symposium, Yokohama, Japan, October 1992.

[08] J. Behein, "Security First in Europe's Mobile Communications," *Telecom Report International,* Vol. 17, No. 1, 1994.

[09] F. Mademann, "General Packet Radio Service—A Packet Mode Service Within the GSM," International Switching Symposium, Berlin, April 1995.

[10] J. Hamalainen, "High Speed Data Service in GSM," ITU Telecommunications Forum 95, Geneva, October 1995.

[11] I. Sharp, "Flexible Service Creation in Mobile Networks," ITU Telecommunications Forum 95, Geneva, October 1995.

[12] TIA/EIA IS-54B, "Cellular System Dual-Mode Mobile Station–Base Station Compatibility Standard," Telecommunication Industries Association, Washington, DC, 1992.

[13] TIA/EIA IS-136.1, "800 MHz TDMA Cellular–Radio Interface-Mobile Station–Base Station Compatibility–Digital Control Channel," Telecommunication Industries Association, Washington, DC, 1994.

[14] TIA/EIA IS-41C, "Cellular Radiocommunications Intersystem Operations," Telecommunication Industries Association, Washington, DC, 1995.

[15] R. H. Glitho, "Use of SS7 in D-AMPS-Based PCS: Orthodoxy vs. Heterodoxy," *Personal Communications Magazine,* Vol. 4, No. 5, June 1997.

[16] N. Nakajima, "Japanese Digital Cellular Radio, in *Cellular Radio Systems,* D. M. Balston and R. C. V. Macario, eds., Artech House, Boston, 1993.

[17] K. Minoru, K. Kinoshita, M. Eguchi, and N. Nakajima, "An Outline of Digital Communication Systems," *NTT Review,* Vol. 4, No. 1, January 1992.

[18] Research & Development Center for Radio Systems (RCR) Japan, RCR STD-27, Personal Digital Cellular Telecommunication System, RCR, Tokyo, 1991.

[19] K. Kinoshita, K: Minoru, and N. Nakajima, "Development of a TDMA Digital Cellular System Based on Japanese Standards," IEEE Vehicular Technology Conference, St Louis, May 1991.

[20] A. H. M. Ross and K. S. Gilhousen, "CDMA Technology and the IS-95 North American Standard," in *The Mobile Communications Handbook,* Jerry D. Gibson, ed., CRC Press, Boca Raton, FL, and IEEE Press, Piscataway, NJ, 1996.

[21] TIA/EIA IS-95A, "Mobile Station–Base Station Compatibility Standard for Dual-Mode Wideband Spread Spectrum Cellular System, Telecommunication Industries Association, TIA, Washington, DC, 1995.

[22] "Code-Division Multiple Access," Chapter 13, in *Spread Spectrum Communications Handbook,* M. K. Simon, J. K. Omura, R. A. Scholtz, and B. K. Levitt, eds., McGraw-Hill, New York, 1996.

[23] K. S. Gilhousen, I. M. Jacobs, R. Padovani, A. J. Viterbi, L. A. Weaver, and C. E. Wheatley, "On the Capacity of a Cellular CDMA System," *IEEE Transactions on Vehicular Technology*, Vol. 40, No. 2, May 1991.

[24] A. J. Viterbi, "The Path to Next Generation Services with CDMA," 1998 CDMA Americas Congress, Los Angeles, November 17–20, 1998.

[25] A. K. Kriplani, "Development of 3G Standards in the Americas," 1998 CDMA Americas Congress, Los Angeles, November 17–20, 1998.

CHAPTER 4

LOW POWER WIRELESS COMMUNICATIONS SYSTEMS AND NORTH AMERICAN PCS

4.1 BACKGROUND

Analog cordless telephones are in common use in residential applications, where the telephone cord is replaced by a wireless link to provide terminal mobility to the user within a limited radio coverage area. Digital cordless telecommunication systems are intended to provide terminal mobility in residential, business, and public access applications where the users can originate and receive calls on their portable terminals as they change locations and move about the coverage area at pedestrian speeds. It is also anticipated that the same terminal can be used in the three application environments: at the residence, at the workplace, and at such public locations as airports, train and bus stations, and shopping centers.

In contrast to cellular radio, cordless telecommunications standards primarily offer an access technology rather than fully specified radio access and network standards. Cordless terminals generally transmit at lower power than cellular, resulting in the use of microcells. In high density (in-building) applications much smaller cells (picocells) may be used so that significantly higher traffic densities can be supported. Further, cellular networks generally operate in an environment characterized by regulation (which may vary from country to country), a limited number of operators, centrally managed frequency resources, and relatively high infrastructure costs. Cordless telecommunication systems, on the other hand, are expected to operate in an unregulated, open-market environment, where system installation and (frequency) planning cannot be coordinated or planned, and cost and performance as perceived by the end user are key factors in market acceptance.

The cordless telephony standards adopted by the European Telecommunications Standards Institute (ETSI) are CT2 (Cordless Telephony 2) and DECT (Digital Enhanced Cordless Telecommunications, previously known as

Table 4.1 Radio Specifications for Cordless Telecommunication Systems

Radio Parameter	CT2	DECT	PACS	PHS
Access method	FDMA/TDD	TDMA/TDD	TDMA/FDD	TDMA/TDD
Spectrum allocation	864-868 MHz	1880-1900 MHz	U.S. PCS band	1895-1918 MHz
Carrier spacing	100 kHz	1728 kHz	300 kHz	300 kHz
Number of carriers	40	10	16 pairs/10 MHz	77
Channels/carrier	1	12	8/pair	4
Modulation	GFSK	GFSK	pi/4 shifted QPSK	pi/4 shifted QPSK
Transmission rate	72 kb/s	1152 kb/s	384 kb/s	384 kb/s
Speech coding	32 kb/s ADPCM	32 kb/s ADPCM	32 kb/s ADPCM	32 kb/s ADPCM
Frame duration	2 ms	10 ms	2.5 ms	5 ms
Peak output power	10 mW	250 mW	200 mW	80 mW

Digital European Cordless Telecommunications). PHS (Personal Handy-phone System) was standardized by the Telecommunication Technology Committee (TTC) in Japan, and PACS (personal access communication system) is a North American low power (or "low tier") PCS standard primarily based on the wireless access communication system (WACS) developed by the Bell Communications Research (Bellcore) laboratories; PACS also includes some aspects of PHS. All these systems are digital systems and can be used in residential cordless, wireless PABX, and low mobility (pedestrian speed) public and private (in the licensed and unlicensed frequency bands) applications. Systems based on these standards are being deployed at a rapid rate in Europe, North America, Japan, and Southeast Asia.

In the United States, a single standard for cordless business, residential, and public access applications is not likely. However, standards based on CT2, DECT, PHS, and PACS radio interfaces have been developed to address these markets in the licensed and unlicensed 1900 MHz PCS bands. The U.S. standards in the unlicensed band category include PCI (personal communications interface) based on CT2, WCPE (wireless customer premises equipment) based on DECT, PACS-WUPE (wireless user premises equipment) based on PHS, and PACS-UB (based on WACS). Table 4.1 summarizes the radio characteristics of CT2, DECT, PACS, and PHS, the four main standards on low power wireless or cordless telecommunications.

4.2   CT2 (CORDLESS TELEPHONY 2) SYSTEMS

4.2.1   Introduction

CT2 is a second-generation cordless telecommunication system—a digital cordless standard designed to be used for wireless PABX and key systems, cordless residential telephones, and public cordless access (telepoint) services. The common air interface (CAI) for CT2 specifies the open standard for the radio path and ensures interoperability of terminals and base stations from different vendors. The CT2 standard originated in the United Kingdom for providing public telepoint services and was adopted in 1992 by ETSI as an interim European Standard for cordless telecommunications (I-ETS 300 131). A number of systems for business and public cordless applications based on the CT2 standard are in operation in countries around the world.

Although the CT2 standard has supported two-way calling from the beginning in all environments including public cordless access, the original networks in which public cordless or telepoint service was implemented did not use this feature. However, newer public cordless implementations are taking advantage of such basic CT2 features as two-way calling and intercell handoff, as well as new features like advanced location tracking now available in the new edition of the CT2 standard (I-ETS 300 131, edition 2).

The CT2plus standard adopted by Canada is fully compatible with CT2 and provides as standard features more efficient location tracking mechanisms, and faster incoming call delivery and handoff. Because CT2plus offers the combination of more RF channels and their efficient allocation via beacon channels, dense usage areas such as large, high density office and business environments are able to accommodate significant increases in their traffic capacities.

The main application environments for CT2 include residential cordless setups, public telepoint systems, and wireless PBX and key systems. In principle, a CT2 handset purchased for use in one environment (e.g., home or work) can also be used in other environments (e.g., in telepoint). In case of telepoint systems, base units are installed in public places like airports, railway and bus stations, shopping centers, convention centers, and sports complexes so that people with CT2 terminals can originate calls (an alternative to pay phones). As shown in Figure 4.1, this application needs some form of authentication procedure to ensure that would-be callers are authorized to use the service, and it also needs a set of procedures for billing purposes. Telepoint therefore

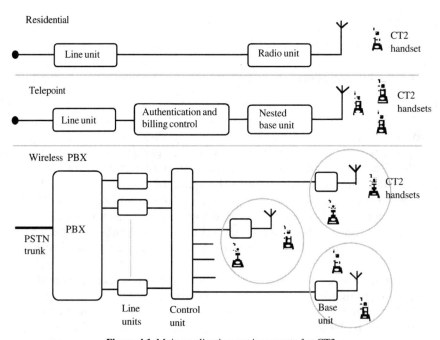

Figure 4.1 Main application environments for CT2.

represents a basic public communications service for low mobility, limited-coverage applications; it does not compete with the high mobility, wide-area coverage of cellular mobile service.

4.2.2 Radio Aspects

The spectrum allocation for CT2 is a 4 MHz slot from 864 to 868 MHz. This is divided into 40 channels, each with 100 kHz. The CAI specifies the modulation as two-level frequency shift keying (FSK) shaped with a Gaussian filter. As shown in Figure 4.2, the binary 0 and 1 are represented by frequency shifts of 14.4 to 25.2 kHz below and above the carrier frequency (Fc), respectively. In the CT2 system, each of the 100 kHz channels is used as a two-way channel in which the transmission and reception of information from the handset (or the base unit) takes place in bursts, which are interleaved on the same channel (radio carrier frequency), as illustrated in Figure 4.3.

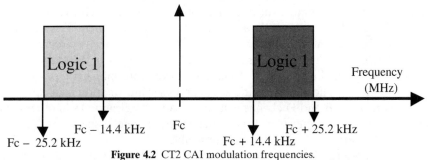

Figure 4.2 CT2 CAI modulation frequencies.

Thus CT2 is an FDMA/TDD system in which the rate of switching between transmit and receive on a given FDMA channel is set at 500 Hz, representing transmit and receive cycles of 1 ms each. In each burst, nominally 72

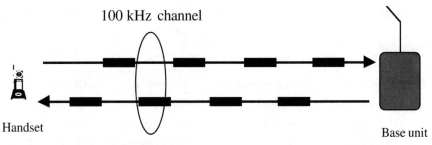

Figure 4.3 Time division duplex (TDD) operation in a CT2 channel.

bits of data are available for speech, control, signaling and base–handset synchronization, leading to an overall data rate of 72 kb/s. The signaling layers 1, 2, and 3 of the CT2 common air interface specification provide the mechanisms necessary to permit speech, control signaling, and synchronization information to be carried efficiently and correctly between the CT2 handset and base unit combinations. The preferred speech codec used in CT2 is based on the ITU-T Recommendation G.721, which is the specification for 32 kb/s ADPCM (adaptive differential pulse code modulation).

4.2.3   Layer 1 Signaling

Signaling layer 1 provides the mechanism for initiation of a two-way digital link over the radio path and the selection of a suitable radio channel for information transfer. Signaling layer 1 also ensures choice of data and signaling channels with minimum interference over the radio path. To this end, a B channel (for speech or data), a D channel (for control and signaling), and a SYNC channel (for bit-and-burst synchronization) are multiplexed together within the 72 kb/s frame structure. The combination in which these channels are multiplexed depends on the requirements to support the different phases of a call. Thus, an information channel may be absent in some circumstances, depending on whether the call is at the link setup stage or in the speech transmission/reception stage. Table 4.2 summarizes the multiplexing types used in CT2 and their applications for link setup and user information transfer.

Whereas the B and D channels have the same application as in other communication environments (e.g., ISDN), the SYNC channel is specific to CT2 and consists of a series of 1s and 0s (i.e., a preamble of $1, 0, 1, 0, 1 \ldots$) followed by special bit patterns (words) that are used to identify a radio channel for link setup. The words delineate time instances within a burst which enable a handset/base unit pair to achieve burst synchronization. Four such bit patterns (SYNC words) are used in CT2 for channel setup initiation and acknowledgment. When a call is being initiated on the radio channel, a handset or a base unit uses channel markers (CHM): the handset will send CHMP (channel marker—portable) while the base unit sends CHMF (channel marker—fixed). When the channel is established, the handset responds with SYNCP and the base unit by SYNCF.

Multiplex 1 is used when the handset and the base units are in the

Table 4.2 Summary of Multiplexing Types in CT2 and Their Applications

Multiplex Type	Application	Direction	Channels Used
Multiplex 1	Speech/data traffic	Handset to base	D and B
Multiplex 2	Link setup	Base to handset	D and SYNC
Multiplex 3	Link setup	Handset to base	B and SYNC

speech/data transfer stage in which the link has been established and the handset and the base unit are in bit and burst synchronization. Speech or user data are carried in the B channel, and the base and handset identities are interchanged through the D channel to assess the link integrity. The level of bit errors encountered over the D channel is used to assess the overall error performance of the channel for speech, and if the bit errors on the D channel are too high, a change to a better channel is attempted (while speech transfer is in progress). While D channel errors are corrected by retransmissions, there is no provision for error detection/correction on the B channel.

The frame format for multiplex 1 is shown in Figure 4.4. CT2 provides for two different multiplex 1 types. Multiplex 1.4 (shown in Figure 4.4) provides for 4 bits/burst for the D channel and a *dead time* of 4 bits. The dead time is required for a proper changeover from the transmit to receive mode and vice versa. In the case of multiplex 1.2, the D-channel bits are reduced to 2 bits/burst, with a 6-bit dead time. Though the B-channel rate in both cases remains at 64 bits/burst (32 kb/s), the D-channel rate for multiplex 1.4 is twice that in multiplex 1.2. The decision on whether to use multiplex 1.2 or 1.4 for speech phase is made at the end of link setup phase by interchange of messages in the D channel in multiplex 2 format.

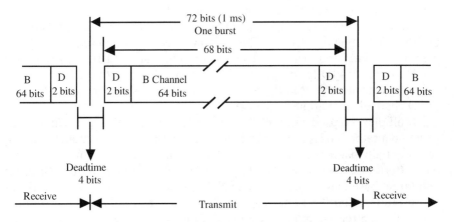

Figure 4.4 Multiplex 1.4 format in CT2.

Multiplex 2 is used for the link setup phase when the base station needs to set up a radio link to a specific handset: for example, for an incoming call to a CT2 handset. At this stage, the base unit needs to signal its identity to the called handset and also needs to establish synchronization with the target handset. The base unit uses the D channel for the former and the SYNC channel for the latter. The structure of multiplex 2 frame in CT2 is shown in Figure 4.5.

Within the 66-bit frame, the SYNC channel consists of a 10-bit preamble

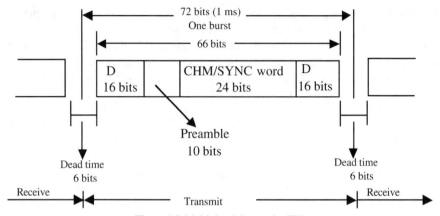

Figure 4.5 Multiplex 2 format in CT2.

(1/0 inversions) and the 24-bit CHM or SYNC word. The 10-bit preamble allows the handset to gain bit synchronization with the base unit at 72 kb/s. To achieve burst synchronization, the CHMF follows the preamble, starting at a fixed point within the burst; when it is detected by the handset, it can then synchronize its receive/send bursts with that of the base unit.

During the link setup phase, the D-channel rate is 16 kb/s (about 8 kb/s after error detection/correction). The D channel is primarily used to carry the base unit and handset identities. The former in the form of LID (link identity) and the latter as PID (portable identity). Having detected the LID, the handset responds with SYNP in the SYNC channel and its own identity (PID) in the D channel, in multiplex 2 format.

Multiplex 3 is used when a CT2 handset wishes to initiate a call and needs to set up a radio link. As shown in Figure 4.6, in multiplex 3 format, the handset essentially pages the base unit for 10 ms with a relatively complex preamble pattern (1/0 inversions), D-channel information containing handset PID and target base unit LID, and the SYNC channel containing CHMP for synchronization. After the 10 ms page, the handset waits for a response from the base unit, in a receive mode, for 4 ms. If no response is received, the handset continues the 10 ms transmit and 4 ms listen cycle in multiplex 3 format for up to 5 seconds.

Since the base unit always operates at the 500 Hz (1 ms) burst rate in multiplex 2 format, repetition of information from the handset (in multiplex 3) is designed to ensure that one of each of the four D and SYNC channels is received at the base unit within the 1 ms base receive window.

When the base has fully decoded the D and SYNC channel parameters, it responds with its own LID and handset PID in the D channel, and SNCF in the SYNC channel, in the multiplex 2 format. The handset detects the SYNCF

Handset: transmits and receives in multiplex 3 format

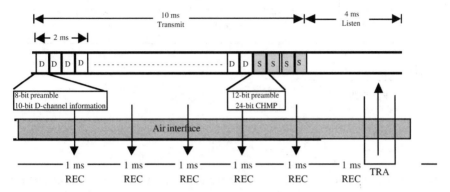

Base unit: transmits (TRA) and receives (REC) in multiplex 2 format

Figure 4.6 Multiplex 3 format in CT2.

in the 4 ms listen cycle in multiplex 3 and, knowing its exact position in the multiplex 2 frame, switches to multiplex 2 in full synchronization with the base unit.

4.2.4   Layer 2 and Layer 3 Signaling

Whereas layer 1 provides the multiplex structures for interchange of information between the handset and the base unit for link setup and speech phase in CT2, layer 2 addresses the control channel protocols for error detection/correction, message acknowledgment, link maintenance, and link end-point identification. This signaling layer facilitates communication of messages between end points without defining their meaning.

Layer 2 packets sent over the radio interface consist of up to six code words. The first code word is a 64 bit address code word (ACW), and subsequent code words contain data in the form of 64-bit data code words (DCWs). Some messages may not require DCWs because they could be accommodated within the ACW. The D-channel message structure for layer 2 messages is shown in Figure 4.7.

In the idle state (no information to be sent), a series of 1/0 inversions is transmitted and, prior to transmission of an information packet, a 16-bit synchronizing word (SYNCD) is sent to the receiving end to indicate that a message is to follow. Both ACW and DCW have 64-bit lengths with an 8 bit/byte format. The ACW and DCW are distinguished by the first bit of the message (first bit = 1 means ACW and first bit = 0 means DCW). The second bit of the ACW defines the format type (FT) of the message. Thus FT bit = 0 implies a fixed-length format for handshake or link end-point identification. FT = 1 is

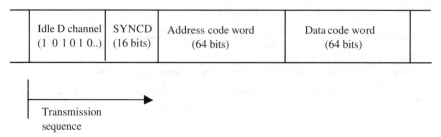

Figure 4.7 D-channel structure for layer 2 messages.

used for variable-length messages for transmission of information at layer 3. FT is significant in only ACWs and is always set to 1 in DCWs. Out of the 8 bytes in the ACW, bytes 1 and 2 are used for control, and bytes 7 and 8 for CRC and parity. Bytes 3 to 6 are available for information transfer.

Layer 3 signaling ensures error-free delivery of meaningful telephony messages across the radio link. The messages are defined as a group of information elements, which may refer to the handset keypad, handset display, or PBX access messages. For example, the keypad information elements will consist of digits 0 through 9 as well as * and #. Layer 3 messages in CT2 can be up to 29 bytes in length.

Generally the 40 channels available, combined with the dynamic channel selection procedures used in CT2, are adequate to meet the traffic capacity needs of residential and small business operations. However, frequency reuse techniques need to be deployed to increase the effective capacity to meet the needs of large business (PBX) systems. The multicell coverage techniques used in the cellular systems can also serve for this purpose. In this case, the resulting cell sizes are small (a few meters in diameter), and the cells are three-dimensional. Allowance also needs to be made for signals passing through walls and ceilings. In a large building complex, it is some times difficult to provide good and even coverage by suitable placement of base units and antennas alone. In such cases, radiating cables may be used to provide radio coverage to hard-to-reach corners.

4.3   DECT (DIGITAL ENHANCED CORDLESS TELECOMMUNICATIONS)

4.3.1   Introduction

The European Digital Enhanced Cordless Telecommunications standard was developed by ETSI to serve a range of applications that included the following:

- residential systems
- small business systems (single site, single cell)
- large business systems (multisite, multicell)
- public cordless access systems
- wireless data (LAN) access systems
- evolutionary systems (fixed wireless access, cordless access to cellular)

The DECT standard has a modular structure so that requirements of different applications can be accommodated in a cost-effective manner. For example, data LAN and PABX applications represent closed environments requiring minimum open standards, whereas with an application like public cordless access, it is essential that products from different vendors not only coexist but interwork with each other. The DECT standard therefore includes the public access profile (PAP), a set of specifications to which all DECT equipment designed for public cordless access must conform. Currently, for closed user group applications like PABX and LANs, there is no requirement to support PAP. However, during 1993, the European Commission and ETSI decided to extend the concept of PAP to cover a minimum level of mandatory conformity for both residential and PABX applications. ETSI has therefore developed the generic access profile (GAP), which will provide cordless basic voice service as people migrate with their GAP-compatible handsets across different application environments. Another profile recently specified by ETSI is related to DECT/GSM interworking and is called the GSM interface profile (GIP). The DECT standard can support high local user density and is suitable for operation in unpredictable and highly variable traffic and propagation conditions. As mentioned earlier, both DECT and CT2 are primarily radio interface specifications. However, the DECT standard also includes a network architecture without specifying all the necessary procedures and protocols.

The data and multimedia capabilities for DECT have recently been enhanced by such newly developed DECT standards as DECT packet radio service (DPRS) and multi media access profile (MMAP). A DECT-based radio transmission technology (RTT) proposal has also been submitted to the ITU-R TG8/1 as a candidate low power wireless radio interface for IMT-2000.

A functional or conceptual view for a DECT system is shown in Figure 4.8. The DECT system is composed of a number of functional entities, including the following:

- the DECT portable handset or terminal, which may include a cordless terminal adapter (CTA) to support more general applications (e.g., fax, data)

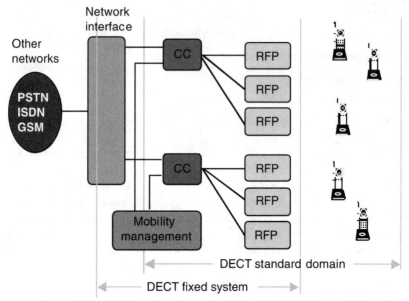

Figure 4.8 Functional or conceptual model for DECT.

- the DECT radio fixed part (RFP), which supports the physical layer of the DECT common air interface

- the cordless cluster controller (CC), which supports the MAC, DLC, and network layers for one or more clusters of RFPs and, in a multicell environment, handles intercell handoffs

- the network-specific interface unit (IU) or interworking unit provides required interfaces to connect to specific networks (PSTN, ISDN, GSM etc.) and, in a multicell environment, may also provide call/connection control for DECT terminals

- the mobility management function, which supports centralized authentication for public access or telepoint applications or other mobility management functions in multilocation PBX-based networks

4.3.2 Radio Aspects

In Europe the 1880–1900 MHz band has been set aside for DECT. To utilize the available 20 MHz band in an efficient and flexible manner for supporting voice and data applications, the DECT standard provides for space, frequency, and time distribution. Space dispersion in DECT is supported through the frequency reuse feature based on the cellular concept. To provide frequency distribution, the available spectrum is segmented into 10 carrier frequencies (frequency channels) from 1881.792 to 1897.344 MHz with a separation of 1.728 MHz [i.e., by deploying frequency division multiple access (FDMA)]. Time distribution is achieved by using time division multiple access (TDMA) and time division duplex (TDD) methods, whereby each frequency channel supports 12 duplex time slots or 32 kb/s channels. The key radio parameters associated with DECT were summarized in Table 4.1; the FDMA/TDMA/TDD structure of DECT channels is illustrated in Figure 4.9.

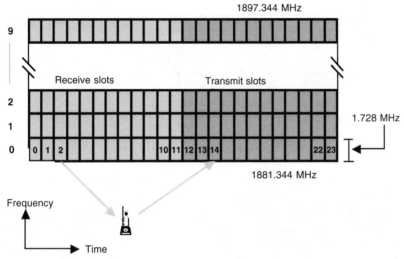

Figure 4.9 DECT radio channel structure.

Thus a total of 120 channels, each with a 32 kb/s data rate, are available in this FDMA/TDMA/TDD structure to carry voice and data traffic; additional capacity required for business systems is provided by invoking frequency reuse in a cellular configuration. In the cellular configuration used in business systems and telepoint, DECT also supports handoffs across cells. The multiplex burst structure used in DECT for carrying voice and signaling information is shown in Figure 4.10.

The 10 ms frame is partitioned into a total of 24 time slots: 12 for fixed-to-portable transmissions and 12 for portable-to-fixed transmissions, so that a normal telephony call uses two matched time slots. The physical layer (PHL) packet consists of a 32-bit sync signal followed by 388 bits of media access layer (MAC) data. Within the MAC packet the 48-bit signaling and CRC field is divided into 16 bits for the CRC, which is preceded by 48 bits that are used for signaling or control (C), the beacon or broadcast (Q), and the paging (P) channels. Of the remaining 324 bits, 320 bits are used for traffic information (I) at 32 kb/s and the last 4 bits (optionally extensible to 8 bits) are used to detect

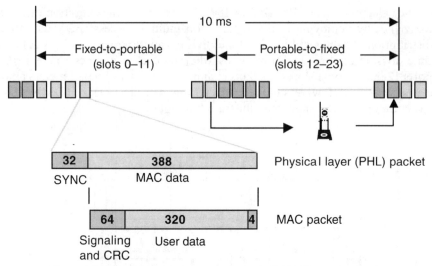

Figure 4.10 DECT Multiplexing structure for individual bursts (packets).

collisions between packets (bursts) from independent or unsynchronized systems. The multiplex structure in Figure 4.10 is specific to the transfer of user information. DECT uses a different channel structure for the call setup phase.

Though the basic data rate for DECT is 32 kb/s, a number of time slots (not necessarily adjacent or in the same frequency channel) can be concatenated to provide higher bit rate channels (e.g., 64 kb/s using 2 time slots and 384 kb/s using 12 time slots are defined in DECT for ISDN and LAN applications, respectively). Higher capacity channels can be set up in both the simplex and duplex modes. The DECT standard also defines half-rate channels (using half time slots), which can be used for deployment of low bit rate codecs.

The mechanism for channel selection in DECT is known as continuous dynamic channel selection (CDCS), whereby channels are chosen on the basis of least interference within all the available DECT channels (any time slot on any frequency carrier). When a connection is needed, the portable selects a channel with least interference (highest quality), and continues to scan the remaining channels. If a better channel becomes available during the call, the portable switches to this channel. The old and the new connections can overlap in time, permitting seamless handoff.

4.3.3   DECT Radio Link: -Layered Architecture

DECT supports a diverse set of services and applications that requires a clear distinction between the processes that are part of the DECT specification and those that the DECT network simply supports. This in turn calls for a struc-

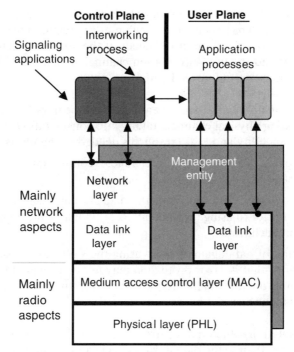

Figure 4.11 DECT radio link: layered architecture.

tured protocol architecture based on OSI principles. The DECT protocol architecture, which follows the OSI layering, is shown in Figure 4.11.

4.3.3.1 Physical Layer. The DECT physical layer (PHL) supports the following functions:

- modulation and demodulation of the radio carriers with the bit stream rates shown in Figures 4.9 and 4.10
- creation of physical channels with fixed throughputs
- monitoring of the radio environment to

    activate physical channels on request by the MAC

    recognize attempts for set up of a physical channel

    acquire and maintain synchronization between transmit and receive channels

    notify the "management entity" about the status of the physical channel (e.g., quality)

For details of the physical channels, see Table 4.1, and Figures 4.9 and 4.10.

4.3.3.2 MAC Layer. The primary task of the MAC layer is to allocate radio resources by dynamically activating and deactivating physical channels and by optimum consideration of the needs of the associated signaling channel (C channel) and the user information channel (I channel). The following classes of functions are supported by MAC:

- creation, maintenance, and release of radio bearers by activating and deactivating physical channels, taking into consideration the changes in bearer capability and radio resources needed for handoffs

- multiplexing of the four logical channels (control, information, paging, and broadcast) onto the physical channel

- segmentation of data frames or the data stream received from the upper layer prior to using the data bursts provided by the physical layer to transport such frames

- protection of data against errors using the CRC and retransmission if necessary—that is, error protection may *not* be provided for user data, depending on the quality requirements for the service (e.g., speech transmissions may not be error-protected).

It should be noted that the error control function in DECT is assigned to the MAC layer rather than to the data link control layer (normally tasked with error control) because this function is best performed on a (radio) link-by-link basis, which in DECT exists only up to the MAC layer.

4.3.3.4 Management Entity (MGE). The management entity (MGE) is responsible for a number of functions in DECT that involve only one side of the communication link and as such do not appear in an OSI layer. These include such functions as radio resource control (choice of free channels, assessment of channel quality), mobility management (registration of DECT portables), and error handling (call termination at radio link interruption).

The data link layer (DLC) and the network layer in the foregoing model are considered in Section 4.3.4 on the network aspects of DECT.

4.3.4 DECT Network Aspects

A primary object of the DECT standard has been to ensure maximum applicability of the standard in a variety of environments based on existing or emerging public and private networks. Thus the DECT air interface and the related radio link protocol architecture were defined so that DECT users can access a range of networks, particularly:

- analog PSTN, PBXs and key systems
- digital ISDN public networks, PBXs and key systems

- GSM networks
- telepoint networks
- local-area networks
- public X.25 and frame relay networks (PSPDN)

As shown in Figure 4.8, the flexibility in the DECT system is provided by the interface or interworking unit, which allows the DECT terminals to connect to a range of diverse public and private networks. The radio link protocol architecture (Figure 4.11) provides the necessary hooks to facilitate such interworking with external networks (at the control level as well as at the user information level) in the form of service access points (SAPs) to signaling applications, interworking processes, and application processes. A more detailed view of the data link and network layer protocols is shown in Figure 4.12.

4.3.4.1   Data Link Control Layer (DLC) Protocol.   The primary role of the data link protocol is to create and maintain reliable connections between a DECT portable and the central system. Separate datalink protocols

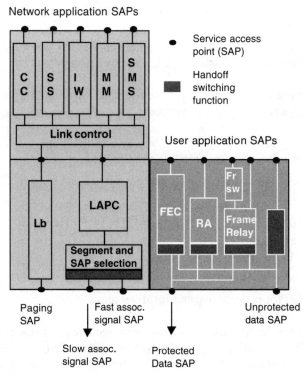

Figure 4.12 Overview of data link and network layer protocols.

are defined for the C plane and the U plane. The former contains two DLC protocols: Lb and LAPC (link access procedure for C plane). Lb offers a connectionless broadcast service and is intended to support paging of DECT portables for alerting them to incoming calls.

LAPC uses variable-length frames, and its characteristics are matched to the underlying MAC service (e.g., segment sizes are chosen to align with inherent MAC timing boundaries to ensure synchronization). LAPC is used for in-call signaling and is primarily derived from LAPD (Q.921), the ITU-T ISDN layer 2 standard, and also draws from the GSM layer 2 specification. LAPC includes the following features:

- supports point-to-point services only (point-to-multipoint services use Lb)

- uses per-call logical link numbers instead of terminal equipment identities (TEIs), both to accommodate the volatility of TEIs in the DECT environment and because mobility in DECT is supported at the MAC layer as well as at the network layer

- uses variable-length frames (identified by length indicator bits), which permits local optimization of frame overheads against retransmission and handoff delays

- allows long network packets to be carried by several DLC packets, thereby reducing handoff and retransmission delays

The user plane (U-plane) services provided by the data link protocol depend upon the application. Possible services that can be supported include the following:

- null service, offering unmodified MAC bearers

- frame relay service, for simple packet transport (per ITU-T Recommendation I.221)

- frame switching, for time-critical packet transport

- fully error-protected (with FEC), for delay- and error-sensitive applications (e.g., video)

- rate-adapted digital data services

- default for proprietary and future expansions

Simple cordless telephony service does not need any of these protocols in the U plane. The foregoing U-plane services are intended as tools that can be used by the vendors or service providers to tailor and offer value-added services.

4.3.4.2   Network Layer Protocol. The DECT network protocol is defined for the control plane (C plane) only. It provides the means to request, allocate, manage, and deallocate key resources in the central system and the portable. The capabilities of the DECT system in terms of its flexibility to support a wide range of applications and interface to diverse networks is determined by the characteristics of the network layer protocol. The techniques used in designing the network protocol were drawn from the ISDN user–network protocol at layer 3 (ITU-T Recommendation Q.931) and to some extent from the GSM layer 3 protocol. Since, however, the DECT architecture and requirements differ considerably from both ISDN and GSM, DECT network layer protocol is significantly different and new. The DECT network protocol structure is essentially modular and it is intended to support a number of protocol functions. Table 4.3 summarizes the current list of the protocol functions and their capabilities.

Table 4.3 Network Layer Protocol Functions and Their Capabilities

Protocol Function	Function Capabilities
Call control (CC)	Establishment and release of network connections, negotiation of connection capabilities, control of interworking units, and transfer of call-related supplementary services
Supplementary services (SS)	Support of call-independent services (call forward, follow-me, advice of charge, reverse-charging, etc.)
Interworking unit (IW)	Protocol conversion between DECT signaling messages and appropriate protocol messages in the external network
Mobility management (MM)	Management of identities, authentication, location management, and subscription procedures
Short message service (SMS)	Supports SMS service for DECT portables in a DECT/GSM interworking environment
Link control (LC)	Matches and coordinates various logical links provided by the data link layer to the needs of the network layer entities

4.3.5   DECT/GSM Interworking

Besides the application of DECT in the residential, small and large business (key and PBX systems), and public access (telepoint) environments, DECT can provide additional capacity and coverage to GSM service providers. DECT/GSM interworking requirements are captured in the GSM interworking profile (GIP), which includes specifications for the following:

- service description and provision
- protocol descriptions and access mapping for 3.1 kHz speech service

- interconnection of DECT fixed part and the GSM MSC via the A interface
- implementation of GSM phase 2 supplementary services
- implementation of group 3 fax, SMS, and other bearer services

Thus, according to the GIP specification, DECT common control fixed part (CCFP) connects to the GSM network via the A interface. CCFP, which essentially represents the radio exchange that handles traffic from a number of DECT fixed radio parts appears as another base station controller to the GSM MSC. This architecture for DECT/GSM interworking is shown in Figure 4.13.

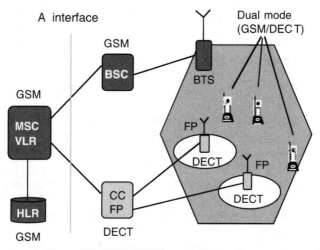

Figure 4.13 DECT/GSM interworking architecture.

The dual-mode terminals deployed in the architecture of Figure 4.13 allow the users to switch to the DECT system when they are in the DECT coverage area. If, however, single-mode terminals (GSM or DECT) are in use, the DECT terminals essentially operate in a centrex mode within the DECT environment, with mobility limited to the DECT coverage area.

4.4 PACS (PERSONAL ACCESS COMMUNICATION SYSTEM)

4.4.1 Introduction

As mentioned earlier, PACS is primarily based on the wireless access communication system (WACS) developed by Bellcore, which till recently was engaged in providing research support for the Regional Bell Operating Companies in the United States. Bellcore's WACS system not only addressed the radio component, but also proposed a network architecture known as the network and operations plan (NOP), to permit the interconnection of WACS radio to a public or private network with intelligent network and mobility management functions. With the allocation of frequency spectrum in the 2 GHz band for personal communication services (PCS) in the United States, multiple radio interfaces for PCS were standardized for operation in the paired licensed band, as well as for operation in the unpaired unlicensed band (Section 4.6).

In essence three versions of PACS have been standardized: one for the licensed PCS frequency band (1850–1910 and 1930–1990 MHz FDD mode) and two for the unlicensed PCS band (1910–1930 MHz TDD mode). Whereas the term "PACS" is applied to the standard for the licensed band, systems for the unlicensed band are known as PACS-UB and PACS-WUPE (wireless user premises equipment); the former is based on the original WACS and the licensed band PACS, and the latter on the Japanese PHS. The important features of PACS include the following:

- a fully integrated networked approach
- can be easily integrated with various cellular systems in a single handset
- can support both private key and public key encryption for authentication and privacy
- downlink preselection receiver antenna diversity and uplink full receiver diversity for better signal quality
- downlink and uplink switched transmit antenna diversity for improving error-free signal transfers
- automatic frequency assignment based on the quasi-static automatic frequency assignment (QSAFA) procedure
- support of subrate (16 and 8 kb/s) and aggregated ($n \times 32$ kb/s) channels
- protocols to support messaging, circuit mode data, packet mode data, and interleaved speech/data services

4.4.2   Functional Architecture for PACS

In one simplified architecture for PACS (Figure 4.14) multiple portable handsets are supported by individual radio ports (RPs) using time division multiple access (TDMA) with frequency division duplex (FDD) operation to distinguish between the uplink and downlink channels. Multiple RPs then subtend on individual radio port control units (RPCUs). The user traffic across the P interface between an RP and an RPCU is separated from the control signals for management of radio functions by using an embedded operations channel (EOC). The RP-to-

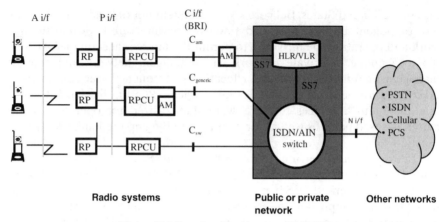

Figure 4.14 Functional architecture for PACS.

RPCU transmission links may take the form of T1 or E1 links, digital subscriber lines (DSL), or high speed digital subscriber lines (HDSL). Whereas the P interface for signaling between the RP and the RPCU is generally a proprietary or provider-specific interface, the C interface between the RPCU and the ISDN/AIN switch is based on the ISDN basic rate interface (BRI).

The radio ports in PACS have a simple design and function primarily as radio frequency (RF) modems. They are powered from the local switch (within 12,000 feet) by means of high speed digital subscriber line (HDSL) technology over copper wires. The RPs can be easily mounted on utility poles or building walls. The RPCU contains the necessary intelligence (electronics) to provide management and control functions for managing radio resources. The access manager (AM), which may be integrated with a RPCU or may operate as a stand-alone unit, includes functions to support such tasks as remote database query, assisting in call setup and delivery, automatic link transfers (ALT) during handoff between two RPCUs, and management of multiple radio ports. The access manager functions may reside in such advanced intelligent network (AIN) elements as a service control point (SCP), an intelligent peripheral (IP), or a switch adjunct. The mobility management functions, necessary databases, and protocols (like IS-41) need to be supported in the public or private network that serves the PACS radio port control units and the access managers.

4.4.3 PACS Radio Aspects

The key radio features for PACS were summarized in Table 4.1. The basic frame for the PACS radio has a duration of 2.5 ms (with 8 time slots/frame) at 400 frames/s. The uplink transmissions from the PACS subscriber unit (SU) utilize time division multiple access (TDMA), and the handset receiver operates in the time division multiplex (TDM) mode for the downlink signals. PACS supports a system broadcast channel (SBC) that may be deployed as one of the following channels:

- an alerting channel (AC) for alerting SUs to incoming calls
- a system information channel (SIC) to broadcast system information (including, e,g., subscriber/terminal identities, relevant timers, protocol parameters)
- a priority request channel (PRC) to be used by SUs for emergency call

The user information (voice or data) is carried in the 80-bit fast channel (FC) or 10-bit slow channel (SC). The frame and burst structure for PACS radio is shown in Figure 4.15.

A time slot in the PACS frame consists of 120 bits, with 80 bits allocated

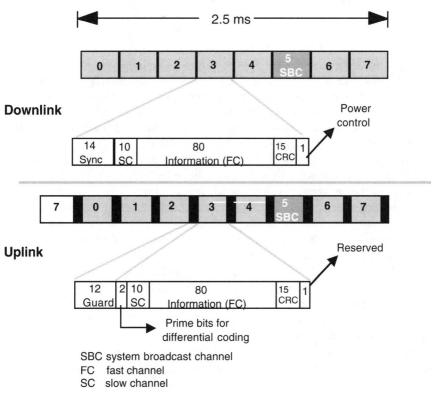

Figure 4.15 Frame structure for PACS.

for the payload (user and signaling traffic) and 40 bits for overhead in the downlink and uplink frames. In the downlink direction the 40 overhead bits are deployed as follows:

- a synchronization channel (14 bits), which provides synchronization
- a slow channel (10 bits), which may be used for transporting additional synchronization patterns, indication of word errors, signaling information, or user data
- a 15-bit cyclic redundancy check (CRC)
- a power bit for optimizing power output at the subscriber unit

In the uplink direction, the initial 12 bits are used as a guard time between consecutive time slots, which are followed by 2 bits for priming differential decoder at the radio ports. The remaining bits provide the slow channel (10 bits), the fast channel (80 bits), and the CRC (15 bits), with remaining bit as reserve bit.

The 80-bit fast channel available once every frame (2.5 ms) translates into a basic data rate of 32 kb/s, which can drive a good quality speech coder. PACS can also provide subrate channels (16 or 8 kb/s) by spacing the bursts every two or four frames, respectively. Higher data rates can be achieved in PACS by aggregating two or more time slots in a frame. The relatively large bandwidth available for the alerting channel (AC) implies almost negligible blocking rates on the alerting channel for call delivery, even when the alerting/registration area (ARA) is very large (e.g., serving > 200,000 users). The use of large ARAs leads to considerably low registration traffic levels in PACS, as well as reduced network signaling between HLR and VLR.

To enhance the signal quality at the subscriber units and radio ports, a PACS radio system deploys two types of diversity technique. The preselection diversity at the subscriber unit takes advantage of the radio port's continuous transmission, which allows the subscriber unit to conduct antenna diversity measurements. Thus the subscriber unit can determine, immediately prior to an incoming burst, which antenna receiver will provide the best signal. Preselection diversity provides an economical, effective, and efficient diversity mechanism for pedestrian speed applications in PACS. As opposed to the preselection antenna diversity used at the SU, the radio port deploys full selection dual-receiver selection diversity, whereby the uplink signals received independently by the two diversity receivers are demodulated and the radio port selects the receiver with the better quality signal.

The preselection and full selection diversity methods just described provide improvements in the general signal quality received at the two ends of the radio interface. PACS also provides for a mechanism that enhances the probability of receiving an error-free burst. This is achieved by deploying switched transmitter antenna diversity both at the subscriber unit and at the radio port. In this method, each end (SU and RP) informs the other if the previous burst was received error free or in error. If the last burst was received in error, the transmitting end switches to the other diversity antenna for the next transmission.

4.4.4   General Systems Features in PACS

As mentioned earlier in this section, the PACS specification includes the radio interface as well as the necessary radio access and network functions to define a complete, implementable system. The three broad categories of functions that need to be supported are radio resource management functions, mobility management functions, and call and service control functions.

Some of the specific functions or features provided in PACS under these categories are described in Sections 4.4.4.1 to 4.4.4.5.

4.4.4.1   Radio Link Maintenance and Measurements.   The PACS radio link maintenance feature is implemented to ensure adequate signal quality during a call. The procedures used for this purpose include automatic link transfer (ALT), time slot transfer (TST), and power control at the subscriber unit. Time slot transfer is a special case of automatic link transfer in which the call is handed off to another time slot within the same radio port. The decisions to invoke an automatic link transfer and the choice of the receiving RP, as well as the decision for a time slot transfer, are made by the SU. The SU output power levels, on the other hand, are controlled by the RPCU, which utilizes the power control procedure to request an SU to adjust its output power level. Invocation and implementation of these prcedures are based on the following RF measurements undertaken by the subscriber unit (when turned on) and the radio ports on every time slot:

- radio signal strength indication (RSSI), which provides a measure of the cochannel interference power and noise
- quality indicator (QI), which provides an indication of signal-to-interference, and signal to-noise ratios, including the effects of dispersion; QI is used in setting the power control bit for raising the SU output power level
- word error indicator (WEI), which provides an indication of occurrence of one or more bit errors (in a time slot) due to radio link degradation; WEI is used by the SU to indicate the uplink performance

4.4.4.2   Automatic Link Transfer (ALT).   In PACS, automatic link transfer or call handoff is initiated and controlled by the SU based on signal strength and quality measurements from the serving RP, as well as a number of candidate RPs suitable for possible handoff. PACS can support the following types of ALT or handoff procedure:

- transfer from one time slot (channel) to another within the same RP i.e., a time slot transfer (TST)
- transfer from one RP to another RP within the same RPCU (intra-RPCU ALT)
- transfer from one RP to another RP; the two RPs under different RPCUs but within the same switch (inter-RPCU ALT)
- transfer from one RP to another RP; the two RPs under different RPCUs and different switches (inter-switch ALT)
- transfer from one RP to another RP; the two RPs under different RPCUs and different access managers or AMs (inter-AM ALT)

The different types of automatic link transfer just listed are transparent to the subscriber unit; that is, the SU uses the same procedure to invoke an ALT request. It is up to the network to recognize which type of ALT procedure is appropriate in a given instance. As an example, an inter-RPCU automatic link transfer is illustrated in Figure 4.16. The steps involved in this type of handoff can be summarized as follows:

1. The subscriber unit requests an ALT by sending a brief signal to the *new* RP-B by briefly interrupting the conversation on the traffic channel.

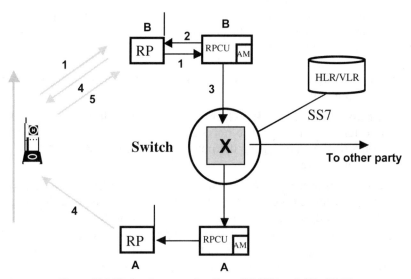

Figure 4.16 Illustrative example of inter-RPCU handoff in PACS.

2. On receipt of the request, the AM associated with RPCU-B transfers the session cipher key to the ciphering unit in the new RP-B.

3. The RPCU-B requests the switch to create a new connection to the SU over the new RPCU-B and RP-B and bridging it to the old path (over RPCU-A and RP-A).

4. The network tells the SU to transfer from the old channel to the new channel through both RP-A and RP-B.

5. The SU confirms the transfer to the new channel, and the network dismantles the old conversation path.

4.4.4.3   Location   Registration/Deregistration.   Location registration (and deregistration) is a key function under mobility management in any mobile network. It allows the network to maintain the current location of a subscriber unit as the subscriber moves around in the network so that incoming calls to the subscriber may be delivered. The registration function is invoked every time the SU is powered on or moves from one registration area (RA) into another, whereby the SU registers with the visited location register (VLR). The VLR then informs the subscriber unit's home location register (HLR) about its current address. In case of registration due to registration area change the VLR on which the SU was previously registered may be informed about its new location to permit updating of records. For incoming call delivery to the SU, its HLR is first consulted about its location, which in turn requests the serving VLR for a routing number/address.

Whereas the registration procedure is similar to that used in cellular or PCS systems that utilize ANSI-41 networking protocols, the deregistration procedure in PACS is somewhat different. ANSI-41 utilizes an *explicit* deregistration procedure: when an SU registers with a new VLR, the previous VLR is always informed about the move so that it may deregister the SU in its records. To reduce signaling load in the network, PACS suggests the use of deregistration by time-out or by polling. In the former case, the SU is deregistered automatically if the SU does not request reregistration within a specific time period since its last request. In the latter case (generally used in PACS-UB), the network periodically sends an alerting message to the SU and, if the SU does not respond to the polling signal within a time-out period, it is deregistered.

4.4.4.4   Authentication and Ciphering.   To prevent fraudulent use of the terminal and to maintain privacy of information over the radio traffic channel, PACS provides for explicit user authentication and ciphering over the radio link. The authentication and key agreement (AKA) protocol in PACS can support a security mechanism based on either a *private key* or a *public key*. The

SU authentication in PACS requires that a unique identity be assigned to an SU (similar to mobile identity number in IS-136 systems or international mobile subscriber identity in GSM). The authentication procedure based on a private key is similar to that used in ANSI-41 systems described in Section 3.3.5.

For radio link ciphering, an algorithm that conforms to the U.S. Data Encryption Standard may be used on the PACS traffic channel. For a newly established channel, the required session key is generated by the authentication and key agreement (AKA) protocol. PACS supports a *security menu* concept, which provides the flexibility to use different options and combinations of AKA procedures and link ciphering algorithms. The security menu is implemented by means of a 4-octet field in the system information channel (SIC): the first two octets indicate the available AKA procedures and the latter two the link encipherment algorithms and modes of operation. As mentioned earlier, during an automatic link transfer or handoff of a call in progress, the session key is transferred to the new serving RPCU for the duration of the call or until the next handoff.

4.4.4.5 Call Origination and Delivery.   The call and service control functions are required to support call originations from the SU, call delivery to the SU, and support of supplemental services or vertical features (e.g., three-way calling, call waiting, calling line identity). Whereas the call origination feature in PACS is similar to that in a mobile network utilizing ANSI-41 procedures, the call delivery feature in PACS follows a procedure that is more efficient in the use of network resources. Whereas in the ANSI-41-based call delivery procedures, the VLR returns the routing address to the HLR on request, in PACS the VLR first pages the SU, and a routing address is returned to the HLR only if the SU is not turned off. Thus, if the SU is turned off, the call will not be routed through the network and it may be given a different treatment (announcement, transfer to a voice mail box) based on the called party's profile.

4.5 PHS (PERSONAL HANDYPHONE SYSTEM)

4.5.1 Introduction

The Japanese PHS (Personal Handyphone System) standard for cordless telecommunications was completed at the end of 1993 by the Research and Development Center for Radio Systems (RCR). The PHS addresses application environments similar to those of CT2, DECT, and PACS (e.g., residential, business, and public cordless access), using a single low power handyphone. The PHS public access service was introduced in Japan in 1994. Besides the basic telephony service, PHS provides data services, including a recently introduced 32 kb/s high speed data service.

The spectrum allocation for PHS is in the 1895–1918.1 MHz band, which is partitioned into 77 carrier frequencies with a separation of 300 kHz. Each carrier is then time division multiplexed into two groups of four time slots, operating in a time division duplex (TDD) mode to provide four duplex channels per carrier. Forty of the 77 carriers (1906.1–1918.1 MHz) are allocated for public systems, and the remaining 37 (1895–1906.1 MHz) to home/office applications. Like other cordless telecommunication systems such as CT2 and DECT, PHS deploys dynamic channel assignment, whereby channel selection is autonomous based on measured signal strength. Various radio parameters were summarized in Table 4.1, along with those for CT2, DECT, and PACS. Features associated with PHS include the following:

- error detection using CRC (PHS does not provide error correction)
- dedicated control channels
- handoff (as an option) at walking speeds
- transmission diversity on the forward link (base to handset)
- group 3 fax at 2.4–4.8 kb/s
- full duplex modem transmission at 2.4–9.6 kb/s and recently at 32 kb/s
- terminals with standby times up to 800 hours

4.5.2   PHS Radio Aspects

As mentioned earlier, the radio access for the PHS is based on TDMA/TDD, where each frequency carrier provides four duplex time slots. PHS standard currently uses a codec based on the ITU-T standard for 32 kb/s ADPCM, with possibility of using a half-rate codec (16 kb/s) or quarter-rate codec (8 kb/s) in the future. The 1895–1981 MHz band allocated to PHS is structured into 77 carriers with 300 kHz separation as shown in Figure 4.17. The 77 carriers are divided between control and communications carriers and also designated for public, private, or common applications.

The signal represented by each carrier is time division multiplexed into eight time slots: four for downlink (CS-to-PS) and four for uplink (PS-to-CS) transmissions, to provide four duplex channels in time division duplex (TDD) operation, as shown in Figure 4.18.

The channel structure for PHS is shown schematically in Figure 4.19; the channels themselves are defined as follows:

CCH (common control channels): used to transfer control and signaling information.

TCH (traffic channels): bidirectional, point-to-point channels used for transfer of user information.

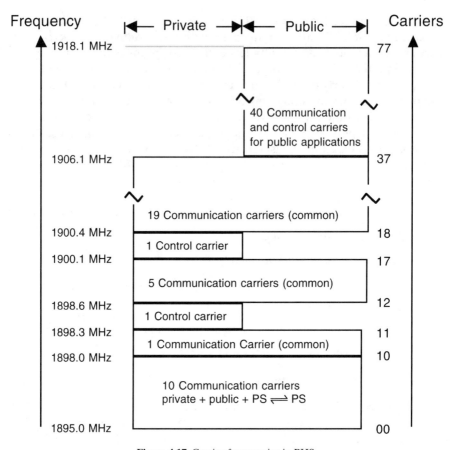

Figure 4.17 Carrier frequencies in PHS.

BCCH (broadcast control channel): a one-way downlink channel for broadcasting control information from the cell station (CS) to the personal station (PS); the BCCH carries channel structure, system information, and related information.

CCCH (common control channel): used for transmitting control information for call setup and connection by means of a paging channel (PCH) and a signaling control channel (SCCH).

PCH (paging control channel): a one-way, downlink channel (CS-to-PS) that transmits the relevant PS and CS identities to the paging area for setting up incoming calls to a PS.

SCCH (signaling control channel): a bidirectional, point-to-point channel that transmits information needed for call connection between a CS and a PS.

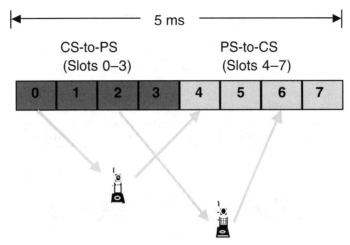

Figure 4.18 The TDMA/TDD time slot arrangement for PHS.

UPCH (user packet channel): a bidirectional, point-to-multipoint channel used for transmission of control signal information and packet data.

ACCH (associated control channel): a bidirectional channel that is associated with the TCH (traffic channel); it carries control information and user packet data needed for call connection. The ACCH, which is ordinarily associated with a traffic channel, is defined as slow ACCH (SACCH). The ACCH that temporarily steals time slots from the TCH is defined as fast ACCH (FACCH).

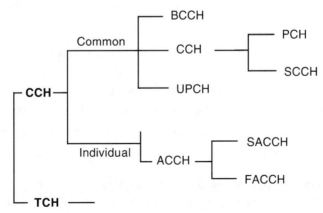

Figure 4.19 PHS channel structure.

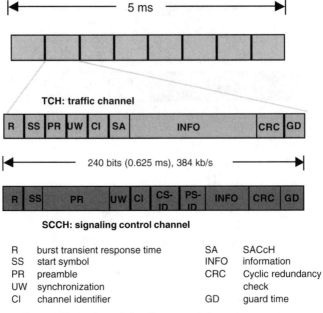

R   burst transient response time   SA   SACcH
SS   start symbol   INFO   information
PR   preamble   CRC   Cyclic redundancy
UW   synchronization   check
CI   channel identifier   GD   guard time

Figure 4.20 Traffic and signaling control channel structures..

Depending on the type of channel, the 240 bits in the frame are allocated to various functions required. Examples of bit distributions for a traffic channel (TCH) and a signaling control channel (SCCH) are shown in Figure 4.20.

4.5.3   PHS Network and Protocol Aspects

4.5.3.1   Functional Requirements.   The general requirements for providing uninterrupted and consistent service to moving PHS stations (PS) are as follows:

- providing subscribed services to a PHS user regardless of PS location
- charging to the same PS based on the distance between the called and calling parties based on current PS location
- maintaining location information for all the PS in the system
- routing and delivery of incoming calls to the current location of the PS
- authentication of the PS as required by the PHS service provider

Support of these requirements is greatly facilitated by the inherent (signaling) capabilities available in digital networks, and by the IN capabilities that are rapidly being introduced. Some of the capabilities required from the signaling protocol for the I interface between the PHS cell station and the network are as follows:

- handling of both voice and data communications
- support of mobility features (registration, authentication, incoming call delivery)
- maintaining independence between CS and the digital network
- support of required supplementary services
- future proof against addition of new services and functions
- flexibility to support multiple PHS operators

The functions supported by the cell station (CS) in the PHS fall into five categories:

1. *Radio access control.* This function is used to manage and administer the radio resources (channels) in terms of connection setup, supervision, and channel quality assessment.

2. *Circuit connection control.* This function provides the capability for setup, release, and administration of circuits between the CS and the digital network.

3. *Location registration.* This is a non-call-associated function that provides the capability for location registration of PHS stations (PS), thus facilitating delivery of incoming calls to a called PHS station.

4. *Supplementary services.* This function enables supplementary services to be requested and provided. In the PHS system, the authentication function is supported as a supplementary service using necessary enhancements to the CS–digital network interface.

5. *Support of signaling protocol.* This function (or a set of functions) terminates the appropriate CS–digital network protocols so that protocol messages for initiating, carrying, and terminating communications can be handled in a robust and flexible manner.

4.5.3.2   *Deployment Scenarios and the CS–Digital Network Interface.*
For PHS system implementation, it is possible to deploy stand-alone PHS networks where all the mobility management functions (e.g., location registration and authentication) and required databases are provided by the PHS service operator. Some PHS operators in Japan have opted for this approach mainly for security reasons. An alternate, and a more commonly used, deployment scenario is to utilize the capabilities available in the local ISDN network. In this configuration, all the network functions for PHS, including location regis-

tration and authentication, are provided by the local ISDN operator, and only cell stations (CS) are provided by the PHS service operator. This type of configuration not only is more economical but also speeds up the introduction of new PHS service offerings. The PHS operators using this option, pay a call-by-call–based access charge to the local ISDN operator. The charge includes the cost of provision and use of the location registration and authentication database as well as the cost of local exchange ports that support the required CS-to-ISDN interface.

Supporting the latter scenario calls for a standardized interface for interconnecting the PHS cell stations to the public ISDN network. The relevant reference points for interconnection of a CS to the digital network are shown in Figure 4.21. The PHS service provider has the option of supporting the CS-to-digital network interface at reference point X1 or X2. In the former case the network termination is considered to be a part of the digital network and provides some advantages from a network management point of view; it requires no changes to the interface if the transmission medium is upgraded (e.g., from copper to fiber). For the X2 reference point the NT is considered to be a part of the CS, and an interface defined at this reference point is better economically because internal (proprietary) signaling can be used in the CS–NT link before being converted to a standardized signal at the X2 reference.

From the ISDN perspective, reference points X1 and X2 are equivalent to the ISDN reference points T and U, respectively. Thus, the obvious basis for specifying the interface (the I interface) at the X1 or X2 reference points is to use the existing ISDN user–network interface specifications with necessary enhancements to support PHS-specific functions. In other words, the I interface for the interconnection of PHS cell stations to the ISDN network is based on the following existing specifications:

Layer 3   Q.931 and Q.932
Layer 2   Q.920 and Q.921
Layer 1   I.431 (X1 reference point) and G.961 (X2 reference point)

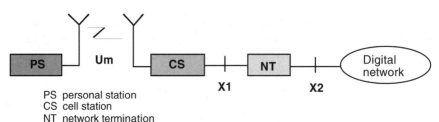

PS personal station
CS cell station
NT network termination

Figure 4.21 Interface reference points for PHS cell station–to-ISDN interface.

The PHS-specific functions that need to be supported at layer 3 (requiring enhancements to basic Q.931 and Q.932 user–network protocols) include:

1. *PS identification.* This function is used to identify the PS so that the digital network can (a) correctly assign call charges for a PS-initiated call (as well as a PS-terminated call if split charging is in use) and (b) deliver an incoming call to the called PS.

2. *Simultaneous sending of an incoming call to multiple interfaces.* This function is used to deliver an incoming call to multiple I interfaces within the location at which the PS is currently registered. This function reduces need for frequent registrations by the PS.

3. *Incoming call during all-circuits-busy state.* This function, is used to send an incoming call message to the PS even when all the circuits between the target CS and the digital network are busy. The PS in this case can try through an alternate CS. This PHS function increases the call completion rate.

4. *Location registration.* This function which is used to register the location of a PS as it moves and to deliver incoming calls to the PS, is provided as a supplementary service via the Q.932 supplementary service procedures.

5. *Authentication.* This function is used to authenticate a PS before ser vice can be provided to the PS in terms of call origination, termina tion, location registration, or handoff. The function is provided via the FACILITY message defined within the Q.932 supplementary service procedures.

6. *Handoff.* This function allows established calls to continue when a PS moves from one radio zone (microcell) to another. The function is supported using the SETUP message in Q.931.

7. *DTMF tone generation.* This function is required because transmission of DTMF tones is not guaranteed on the radio interface. The DTMF signals are sent as numeric data over the D channel, and a FACILITY message in Q.932 is used to invoke the DTMF tone generation function.

Besides the application of PHS in the public (cordless) environment, PHS terminals can also be used in business (wireless PBX) and residential cordless environments. In fact, one of the distinctive features of the PHS system is that a single terminal can be used in all the three application environments. These application environments are illustrated in Figure 4.22 along with the type of interfaces required for interconnecting to the public ISDN network. The I in-

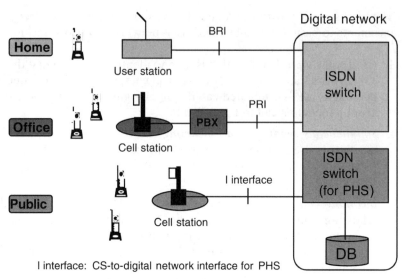

Figure 4.22 PHS application environments.

terface for CS-ISDN interconnection has been defined in the Japanese JT Q.931-b standard.

4.6 PCS IN NORTH AMERICA

4.6.1 Introduction

Allocation of frequency spectrum in the 2 GHz band by the U.S. Federal Communications Commission (FCC) spurred the standardization activity on personal communication service (PCS) in North America in the early 1990s. The FCC has defined PCS as follows: "radio communications that encompass mobile and ancillary fixed communications that provide services to individuals and businesses and can be integrated with a variety of competing networks. "Further, PCS is characterized by the FCC as providing" a broad range of new radio communication services that will free individuals from the limitations of wireline public switched telephone networks and enable individuals to communicate when they are away from their home or office telephones" [24], [25].

The scope of PCS therefore covers high mobility and large cell applications (*high tier*) currently provided by cellular mobile systems like GSM and IS-95, as well as low mobility and small cell applications that are the focus of low power wireless systems like CT2, DECT, PACS, and PHS (*low tier*).

Unlike regulatory agencies in many countries, the FCC decided to define only a minimum set of rules for PCS operators, creating a fair competitive environment in which the choice of standards, technology, and type of services and serving areas for PCS operation is left to the marketplace. Thus the offerings of several large PCS service providers in the United States feature very different combinations of technology, standards, handset design, voice quality, coverage, roaming capability, and mobility.

4.6.2   Frequency Spectrum Allocation for PCS in the United States

The frequency spectrum identified by the FCC for PCS licensed and unlicensed operation is in the 1850–1990 MHz band. In the licensed band, it provides a total of 60 MHz each for uplink and downlink transmission in the frequency division duplex mode with 80 MHz separation (1850–1910 and 1930–1990 MHz), and a total of 20 MHz (1910–1930 MHz) for systems operating in the unlicensed band subject to the FCC-specified *spectrum etiquette* rules. The intention is that licensed band PCS will fulfill wide-area communications requirements while PCS in the unlicensed bands will target localized and in-building users. The 120 MHz spectrum for the licensed band is divided into three blocks of 30 MHz and another three blocks of 10 MHz, as shown in Figure 4.23. The 20 MHz in the unlicensed band is split into two 10 MHz bands—one for voice (1920–1930 MHz) and another for data (1910–1920 MHz). The unlicensed voice band is further channelized into eight 1.25 MHz segments with smaller subchannels permitted at lower power levels with 50 kHz or greater bandwidth.

The licenses for the 2 GHz PCS spectrum (in the licensed band) were made available to the potential operators by the FCC through an auctioning process. A total of 2070 licenses for PCS were auctioned. Of these licenses, 1968 cover regions called basic trading areas (BTAs), which roughly correspond to metro-

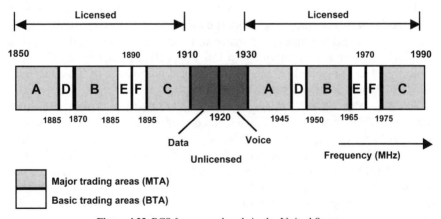

Figure 4.23 PCS frequency bands in the United States.

politan areas or collections of rural counties. These BTA licenses permit use of either 10 or 30 MHz of PCS spectrum. The remaining 102 licenses cover larger regions known as major trading areas (MTAs), which represent a collection of several BTAs. To promote competition, these licenses were distributed over 492 BTAs (four licenses per BTA) and 51 MTAs (two licenses per MTA), and existing cellular operators were restricted to bid for the 10 MHz blocks only.

4.6.3   Radio Interface Standards: PCS Licensed Band

The radio or air interface standards for PCS were developed jointly by ATIS Committee T1P1 and TIA committee TR 46. Whereas most of them were approved as ANSI standards, one is a TIA interim standard and yet another is still at a standards project (SP) stage and remains to be approved as a standard for the licensed band. As expected, there were multiple candidates for the radio interface, but seven were finally approved as radio interfaces for PCS in the licensed band. It was left to the PCS operators who were successful in acquiring the necessary spectrum to decide which radio interface they wish to deploy—their choice was dictated not only by the technical characteristics but also by such factors as cost, speed to market, and compatibility with existing cellular operations.

Five out of the seven standards (GSM, IS-136, IS-95, WACS/PHS, and DECT) were based on existing digital cellular or low power cordless technologies and implementations. The PCS radio interface standards based on these technologies tried to maintain maximum commonality with the existing implementations, and the primary differences from the base standards were generally in the physical and lowest layers, to allow efficient operation in the 1900 MHz PCS band. These digital cellular systems were described in some detail in Chapter 3, and this chapter covers low power cordless systems. The two remaining PCS radio interfaces were new. One is based on a composite CDMA/TDMA technology developed by the Omnipoint Corporation in the United States, and the other on a wideband CDMA technology developed by the OKI Corporation of Japan and a U.S. firm called Inter Digital. The seven radio interfaces specified by the joint technical committee of TIA TR46 and AT15. T1P1 are described briefly as follows.

PCS-1: TIA IS-661 (Composite CDMA/TDMA).   This represents a composite CDMA/TDMA radio interface that utilizes a 2 MHz bandwidth with time division duplexing. It can be deployed in applications for both large cell licensed bands and small cell unlicensed bands. It uses CDMA between cells and TDMA within cells. This radio interface is based on a new technology developed by Omnipoint and provides a flexible structure of 328 kb/s time slots that can support up to 256 kb/s full-duplex or 512 kb/s half-duplex data rates using time slot aggregation. Absence of equalizers further augments the capacity gain provided by CDMA multiple access. The use of TDD obviates the need to clear both uplink and downlink spectra for easy and early deployment. Based on this new technology, Omnipoint was able to acquire PCS spectrum reserved under the special category of *Pioneer Preference*. However, *currently there are no major commerical implementations of this system*. The features of IS-661 include the following:

- Offers full vehicular mobility in (large) macrocells or (small) microcells ranging in cell sizes from 20 miles to several hundred feet in diameter.

- Unlike other CDMA technologies, cell sizes in this system remain unchanged and do not have to be reduced as the cells become loaded near capacity.

- Can support such features as intelligent call screening, multipart ringing, and high speed (64 kb/s) data services.

- Each RF carrier can operate at 256 kb/s rate in the duplex mode, supporting 32 full-duplex voice channels using a 8 kb/s voice coder and up to 64 voice channels using 4 kb/s vocoder.

- Permits interoperability between the licensed and unlicensed band implementations by means of dual-mode terminals which provide wireless quality and variable data rates.

PCS-2: ANSI J-STD 008. 018. 019 (IS-95 Based CDMA).   This 1.25 MHz CDMA interface standard, which is suitable for high and low mobility licensed band applications, is based on the 800 MHz cellular TIA IS-95 CDMA standard. Higher capacities are achieved through CDMA and through the use of variable-rate coders. Not only is interoperation with the 800 MHz cellular bands possible, but the capability exists to evolve to support higher data rates and ADPCM voice coders. The IS-95 system, which is the basis for this PCS standard, is described in Chapter 3, Section 3.5.

PCS-3: ANSI J-STD 014 (PACS).   The personal access communication system (PACS) is primarily derived from Bellcore's wireless access communication system (WACS) with some aspects included from Japan's Personal Handyphone System (PHS). The system is optimized as a cost-effective system for the pedestrian (low mobility) environment. The specification uses an 8 time slot TDMA air interface with FDD for small cell licensed applications, and TDD for small cell unlicensed applications. The specification supports flexible aggregation of time slots, thereby facilitating support of different data rates. Absence of equalizers provides capacity advantage. The PACS system is described in Section 4.4.

PCS-4: ANSI J-STD 009, 010, 011 (IS-136 Based TDMA).   This 3 time slot TDMA radio interface, based on the IS-54/IS-136 800 MHz cellular system, can support high and low mobility and licensed band applications with varying cell sizes. The specification supports full-rate and half-rate voice coders and will provide interoperation between the 2 GHz band PCS and the 800 MHz band cellular versions. Efforts are under way to enhance the specification to support higher data rates. The IS-136 system, which is the basis for this PCS standard, is described in Chapter 3, Section 3.3.

PCS-5: ANSI J-STD 007A (GSM-Based TDMA). This radio interface standard uses an 8 time slot TDMA and is suitable for high and low mobility and large and small cell licensed band applications. It is closely aligned with the DCS 1800 system in Europe, the 1800 MHz version of the GSM radio interface. The specification has time slot aggregation capabilities for higher data rates and supports multiple voice coders. This system is generally referred as PCS 1900. The GSM/DCS 1800 system, which is the basis for this PCS standard, is described in Chapter 3, Section 3.2.

PCS-6: SP-3614 (DECT-Based TDMA for Licensed Band). The specification is based on the European DECT system and uses 12 time slots with TDD. It is suitable for small cell, licensed band applications. A DECT-based radio interface specification for operation in the unlicensed band is also available. The specification for the unlicensed band has been approved as TIA Interim Standard IS-662. The DECT system is described in Section 4.3.

PCS-7: ANSI-015 (New W-CDMA). This specification represents a 5 MHz CDMA radio interface for large and small cell, licensed band applications and is based on wideband CDMA technologies developed by OKI and Inter Digital. The proposed system utilizes coherent detection both in the forward and reverse links and a stable intereference cancellation technology. It claims to provide toll-quality speech based on 32 kb/s ADPCM coding. It also provides voiceband data service at 14.4 and 64 kb/s unrestricted digital data. The forward link (downlink) supports pilot, sync, broadcast, and traffic channels, while the reverse link (uplink) supports access and traffic channels. *Currently there are no commercial implementations of this technology.*

Some of the key technical parameters for these seven radio interface specifications for 2 GHz PCS in North America are summarized in Table 4.4.

4.6.4 Networking for PCS Licensed Band

A generalized reference architecture for PCS was developed and standardized for PCS by ATIS Committee T1P1 and is shown in Figure 4.24. However, most of the PCS radio interface specifications are based on existing digital cellular mobile systems (IS-136, IS-95, GSM) or low power wireless systems (DECT, WACS/PHS), and thus they can utilize network infrastructures associated with these systems. For example, PCS systems based on IS-136 TDMA and IS-95 CDMA use the ANSI-41 specifications for network interoperation. Similarly, PCS 1900 (the GSM-based system) utilizes the GSM networking protocols for network interoperation. For PCS 1900 to operate in the North American en-

Table 4.4 Technical Characteristics for PCS Radio Interface Specifications

Characteristic	PCS-1	PCS-2	PCS-3	PCS-4	PCS-5	PCS-6	PCS-7
Specifications derived from	(new)	(IS-95)	(PACS)	(IS-136)	(GSM)	(DECT)	(new)
Access method	CDMA/TDMA	CDMA	TDMA	TDMA	TDMA	TDMA	CDMA Wideband
Duplex method	TDD	FDD	FDD	FDD	FDD	TDD	FDD
Bandwidth MHz	2	1.25	0.3	0.03	0.2	.728	5
Bit rate, kb/s	32	8, 13.3	32	7	13	2	32
Channels/carrier	32	20	8	3	8	12	28
Modulation	Cont. PSK	OQPSK/ QPSK	pi/4 d-QPSK	pi/4d-QPSK	GMSK	GFSK	OQPSK/QPSK
Error control	—[a]	FEC	—[a]	FEC	FEC	—[a]	FEC
Maximum average subscriber power mW	—	200	12.5	100	125	20.8	500
Frame size, ms	20	20	2.5	40	4.615	10	—
Time slot size, ms	0.625	—	0.3125	6.7	0.557	0.417	—

[a] Some error detection/correction provided but not FEC.

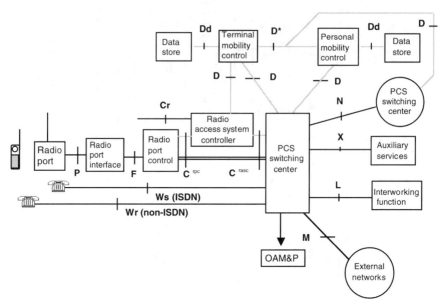

Figure 4.24 Reference architecture for PCS in North America: double lines, bearer interface; solid lines, bearer–signaling interface; dotted lines; signaling interface.

vironment, however, some modifications to the GSM networking protocols were necessary, especially to accommodate differences between the ITU-T SS7 protocols used in the GSM specifications and the North American specifications for SS7 that are used for PCS 1900 networking. ANSI J-STD 023 (PCN-to-PCN intersystem operation based on DCS 1800) addresses networking specifications for PCS 1900. Other PCS systems designed to operate in large cell, high mobility environments have the option of using either the ANSI-41 or the GSM networking protocols.

As described in Sections 4.3 and 4.4, DECT, PACS, and PHS systems do not have a specific network infrastructure requirement. These radio systems directly interface with existing PSTN/ISDN networks via appropriate interface specifications. PCS systems based on DECT or PACS will operate in a similar manner by utilizing the capabilities of the PSTN/ISDN to support small cell, low mobility applications.

4.6.5   Deployment of PCS in the Licensed Band

With the availability of the PCS spectrum and standards, deployment of PCS systems in North America is progressing rapidly. The front-runners in the choice of PCS technology/standards for these implementations in the licensed band seem to be the ones based on existing cellular mobile systems (i.e., GSM-based TDMA, IS-95-based CDMA, and IS-136-based TDMA systems). A major reason for this is the ready availability of technology that ensures reduced *time to market*—a critical factor in a competitive, high demand environment. An additional factor in case of large players with existing cellular networks is interoperability, which facilitates transparent nationwide service for subscribers. As in the case of existing cellular mobile networks, roaming agreements are emerging rapidly between PCS service providers that use similar radio and network technologies.

4.6.6   PCS Standards in the Unlicensed Band

As shown in Figure 4.23, the unlicensed PCS frequency band consists of 20 MHz in the 1910–1930 MHz band divided into two 10 MHz segments: 1910–1920 MHz for asynchronous (data) and 1920–1930 MHz isochronous (voice) devices. The 10 MHz assigned for voice is further divided into eight 1.25 MHz channels. The spectrum *etiquette rules* specified by the Federal Communications Commission impose transmission power levels and emission limits for the devices. The basic concept underlying the etiquette is to limit transmitter power, and duration of transmissions and to provide a "listen before send" protocol so that new transmissions wait for traffic already on a channel.

4.6.6.1   *PACS-UB (Personal Access Communication System—Unlicensed Band).*   As mentioned earlier, PACS-UB is the PACS standard for operation in the unlicensed PCS band in the TDD mode. It is entirely based on the wireless access communication system (WACS) developed by Bellcore modified to conform to the *etiquette rules* set by the FCC for the unlicensed band. The PACS-UB standard was designed to ensure easy interworking and interoperability with PACS (licensed band) by means of dual-mode handsets with minimal incremental complexity. PACS-UB design allows graceful scaling from applications requiring large capacity (e.g., large office wireless centrex systems) to medium-sized PBX and key systems down to residential cordless home ports. Though PACS-UB utilizes TDMA/TDD operation to conform to the unlicensed spectrum allocation and etiquette (as opposed to TDMA/FDD for PACS), there are many common radio and networking features between the two systems.

The PACS-UB also uses 2.5 ms frames with eight bursts per frame. However, for TDD operation the frames are paired (0,2), (1,3), (4,6), and (5,7) to provide four duplex channels per frame or frequency channel. Burst numbers 2, 3, 6, and 7 are used for uplink transmissions (SU to RP) and burst numbers 0, 1, 4, and 5 for downlink transmissions (RP to SU). The 10 MHz unlicensed spectrum is divided into 32 frequency channels, so the PACS-UB radio interface can support a total of 128 traffic channels.

As in the case of PACS, PACS-UB deploys a time slot for system broadcast channel (SBC) for broadcasting system information. SBC is primarily used by the RP in the downlink direction as an alerting channel (AC), an access request channel (ARC), or a system information channel (SIC). In PACS-UB, the system broadcast channel utilizes a superframe (SBC-SF) structure having a duration of 1000 ms. The radio port employs the *basic* SBC-SF for most transmissions, where the *basic* SBC-SF is characterized by two intervals (interval A = 200 ms and interval B = 800 ms); the former is used for alerting and access request, while the latter serves for system information and access request. However, for executing the RF channel access procedure, the radio port utilizes a different superframe structure called *access* SBC-SF, which is divided into four intervals (interval A = 200 ms, interval B = 200 ms, interval C = 400 ms, interval D = 200 ms). While intervals A and B carry information similar to that in the *basic* SBC-SF, interval C is used for searching an acceptable RF channel, and interval C carries access request channel (ARC). The protocol requires that once an RF channel has been successfully identified (during interval C in an *access* SBC-SF), the RP uses the basic SBC-SF for further transmissions. If the transmissions are not acknowledged by the subscriber unit within 30 seconds, the RP relinquishes the RF channel and restarts search for a new RF channel using the *access* SBC-SF. These SBC-SF sequences are shown in Figure 4.25.

As in the case of PACS, PACS-UB utilizes full selection receiver diversity in the uplink direction. In the downlink direction it uses time-delayed transmitter diversity, which takes advantage of a property of TDD operation, namely, that the same frequency is used by both uplink and downlink signals, and thus the radio port can compare the received signal strengths at its two antennas and choose the one with the better signal for downlink transmission.

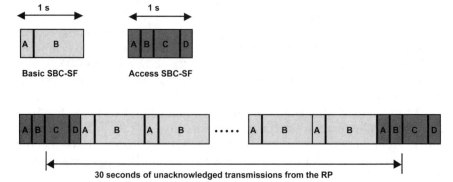

Figure 4.25 SBC-SF sequence for unacknowledged transmissions: PACS-UB.

The performance of this diversity mechanism depends on the delay between the received (uplink) burst and the transmitted (downlink) signal. Generally, the method provides good results for low mobility applications (speeds up to 30 km).

4.6.6.2   *PWT (Personal Wireless Terminal) Interoperability Standard.* PWT is a microcellular, cordless telecommunication system based on the European DECT standard. As in the case of DECT, PWT also operates in the TDMA/TDD mode and is designed as a flexible, cost-effective interface to such applications as wireless PBX, cordless telephony, and wireless local loop. It supports a variety of channels to accommodate voice and data traffic, handoffs, registration, and paging. The reduction to 1.25 MHz channels is dictated by FCC etiquette rules, and it in turn induces change in modulation scheme from GFSK to a narrower band pi/4 DQPSK.

Thus, in terms of the higher layers, PWT is essentially the same as DECT. However, the lower layers [i.e., the physical layer and the medium access control (MAC) layer] have been modified in PWT to make it suitable for operation in the PCS unlicensed band with associated etiquette rules. The following parameters have been modified in PWT specifications:

- frequency band (1910–1930 instead of 1800–1900 MHz)

- carrier spacing (1.25 MHz instead of 1.728 MHz)

- number of carriers (8 instead of 10)

- modulation (pi/4 DQPSK instead of GFSK)

- peak transmit power (90 mW instead of 250 mW)

4.6.6.3   *PCI (Personal Communications Interface) Interoperability Standard.* PCI is a system based on the CT2 standard for operation in the 1920–1930 MHz U.S. PCS unlicensed band. The channel spacing for PCI is incremented by 100 kHz $\times$ n, where n is an integer number between 1 and 99, with the first channel at 1920.10 MHz and the last channel at 1929.90 MHz. Though a maximum of 99 carriers is possible, the FCC allows only 76 carriers to be used. As in the case of CT2, PCI operates in the FDMA/TDD mode. The speech is digitized using the 32 kb/s ADPCM encoding. Time-compressed digitized speech and control information is used to modulate a carrier by means of two-level frequency shift keying (FSK); transmission takes place in 2 ms frames. Each transmitted fram includes one downlink and one uplink burst, thereby avoiding the use of duplex filters in the handset or the base station. The system allows the use of transmission and reception antenna diversity.

The following CT2 parameters have been modified in CTI to conform to the U.S. PCS unlicensed band and the FCC rules:

- frequency band (1920–1930 MHz instead of 864–868 MHz)
- number of usable carriers (76 instead of 40)
- modulation scheme (two-level FSK instead of GSFK)
- frame duration (2 ms instead of 10 ms)
- peak handset transmit power (30 mW instead of 10 mW)

4.6.6.4   PACS-WUPE (PACS—Wireless User Premises Equipment) Standard.   As opposed to PACS-UB, which is a PCS standard for U.S. PCS unlicensed band based on Bellcore's wireless access communication system (WACS), PACS-WUPE is based on the Japanese Personal Handyphone System (PHS), which was aimed at low tier public outdoor service, residential cordless, and PBX applications. The only differences between PACS-WUPE and PHS are primarily in the layer 1 or physical layer specifications, which were required to satisfy spectrum etiquette rules and channel allocation. For example, in PACS-WUPE the number of carriers is reduced from 77 to 32, and the peak output power is reduced from 80 mW to 53 mW.

The key radio parameters for the PCS systems for operation in the U.S. PCS unlicensed band are summarized in Table 4.5.

Table 4.5 Radio Specifications for PCS Systems in the U.S. Unlicensed Band

Parameter	PWT	PCI	PACS-WUPE	PACS-UB
Frequency band, MHz	1920–1930	1920–1930	1920–1930	1920–1930
Multiple access	TDMA/TDD	FDMA/TDD	TDMA/TDD	TDMA/TDD
Carrier spacing, kHz	1250	100	300	300
Number of carriers	8 (or 16)	99	32	32
Bearers/channel	12	1	4	4
Channel bit rate, kb/s	1152 (or 576)	72	384	192
Modulation	pi/4 DQPSK	Two-level FSK	pi/4DQPSK	pi/4QPSK
Speech coding scheme	ADPCM	ADPCM	ADPCM	ADPCM
Speech coding rate, kb/s	32	32	32	32
Frame duration, ms	10	2	5	2.5

4.7 CONCLUDING REMARKS

Cordless telephony or low power wireless systems provide mobile communications in environments where the cellular mobile systems either are not technically suitable or not cost-effective. They fulfill the increasing market demand for applications that need to operate in limited-coverage areas (inbuilding) and limited-mobility situations (walking speeds). Their ability to utilize unlicensed frequency spectrum and the requirements for minimal infrastructure have led to their extensive deployment. Systems like CT2, DECT, PACS, and PHS are gaining rapid market and user acceptance for voice as well as data communications in indoor and outdoor pedestrian environments. The deployment of North American PCS systems (which utilize the 1900 MHz licensed band) is also progressing at a rapid rate—especially for high tier applications in the licensed band and for low tier applications in the unlicensed band.

4.8   REFERENCES

[01] R. S. Swain, "Cordless Telecommunications," Chapter 49, in *Telecommunications Engineer's Handbook*. F. Mazda, ed., London, Butterworth, Heinmann, 1995.

[02] R. S. Swain, "Digital Cordless Telecommunications—CT2," *British Telecommunications Engineering,* Vol. 9, July 1990.

[03] M. W. Evans, "CT2 Common Air Interface," *British Telecommunications Engineering,* Vol. 9, July 1990.

[04] European Telecommunications Standards Institute, "Common Air Interface Specification to Be Used for Interworking Between Cordless Telephone Apparatus in the Frequency Band 864.1 MHz to 868.1 MHz, Including Public Access Services" (ETR 300-131), ETSI, Sophia Antipolis, France, 1991.

[05] European Telecommunications Standards Institute, "Radio Equipment and Systems, Digital Enhanced Cordless Telecommunications (DECT)," Reference Document (ETR 015), ETSI, Sophia Antipolis, France, 1991.

[06] European Telecommunications Standards Institute, "Digital Enhanced Cordless Telecommunications (DECT)," System Description Document (ETR 056), ETSI, Sophia Antipolis, France, 1993.

[07] European Telecommunications Standards Institute, "Digital Enhanced Cordless Telecommunications (DECT)," Common Interface (CI); Part 1: Overview," (ETS 300 175), ETSI, Sophia Antipolis, France, 1992.

[08] European Telecommunications Standards Institute, "Digital Enhanced Cordless Telecommunications (DECT), Public Access Profile (PAP)," (ETS 300 444), ETSI, Sophia Antipolis, France, 1994.

[09] European Telecommunications Standards Institute, "Digital Enhanced Cordless Telecommunications (DECT); Global System for Mobile Communications

(GSM); DECT/GSM Interworking Profile (IWP); Profile Overview," (ETR 341), ETSI, Sophia Antipolis, France, 1996.

[10] European Telecommunications Standards Institute, "Digital Enhanced Cordless Telecommunications (DECT); Data Service Profile (DSP); Profile Overview," (ETR 185), ETSI, Sophia Antipolis, France, 1995.

[11] A. Bud, "System and Network Aspects of DECT," Fourth Nordic Seminar on Digital Mobile Radio Communications, DMR IV, Oslo, June 1990.

[12] M. Hjern, "DECT/GSM Interworking," International Switching Symposium, Berlin, 1996.

[13] S. Kandiyoor, "DECT Meeting Needs and Creating Opportunities for Public Network Operators," International Switching Symposium, Berlin, 1996.

[14] American National Standards Institute, ANSI J-STD 014, "Personal Access Communication Systems Air Interface Standard," ANSI, New York, 1998.

[15] American National Standards Institute, ANSI J-STD 014 supplement B, "Personal Access Communication System Unlicensed (version B)," ANSI, New York, 1998.

[16] A. R. Noerpel et al., "PACS: Personal Access Communications System—A Tutorial," *Personal Communications Magazine,* Vol. 3, No. 3, June 1996.

[17] TIA/EIA IS-41C, "Cellular Radiotelecommunications Intersystem Operations," Telecommunications Industries Association, Washington, DC, 1995.

[18] Research and Development Center for Radio Systems, "Personal Handyphone System, RCR Standard, version 1," RCR STD 28, RCR, Tokyo, December 1993.

[19] H. Okinaka, "A New Advanced Infrastructure for Mobile Communications and Wireless Local Loop Applications," Pacific Telecommunications Conference, Honolulu, January 1996.

[20] S. K. Suzuki et al., 'Personal Handy-phone System Signalling Protocol Architecture—Cell Station–Digital Network," International Conference on Universal Personal Communications, Tokyo, November 1995.

[21] S. Suzuki et al., "Personal Handy-phone System Signalling Protocol Architecture; Study on Inter-Network Interface for PHS Roaming," International Conference on Universal Personal Communications, Tokyo, November 1995.

[22] Y. Hiroshi et al., "PHS Packet Data Communications System," *NTT Review,* Vol. 9, No. 3, May 1997.

[23] J. Segawa and K. Endo, "The PHS Wireless Local Loop," *NIT Review,* Vol. 8, No. 5, September 1996.

[24] Second Report and Order, ET Docket No. 92-9, 7 FCC Red 6886, Washington, DC, 1992.

[25] Memorandum Opinion and Order, GEN Docket No. 90-314, FCC, Washington, DC, June 1994.

[26] C. I. Cook, "Development of Air Interface Standards for PCS," *Personal Communications Magazine,* Vol. 1, No. 4, December 1994 (errata: *Personal Communications Magazine,* Vol. 2, No. 1, February 1995).

[27] A. Fukasawa et al., "Wideband CDMA System for Personal Radio Communications," *Personal Communications Magazine,* Vol. 3, No. 5, October 1996.

[28] American National Standards Institute, ANSI J-STDs 008, 018, 019, "Personal Station–Base Station Compatibility Requirements for 1.8 to 2.0 GHz Code Division Multiple Access (CDMA) Personal Communication System," ANSI, New York, 1996.

[29] American National Standards Institute, ANSI J-STD 007, "Personal Communication Services PCS 1900 Air Interface Specification," ANSI, New York, 1997.

[30] American National Standards Institute, ANSI J-STD 011, 'PCS IS-136 Based Air Interface Compatibility Standard," ANSI, New York, 1996.

[31] American National Standards Institute, ANSI J-STD 023, "PCN to PCN Intersystem Operation Based on PCS 1900," ANSI, New York, 1996.

[32] S. Kwan, "Overview of PCS Unlicensed Wireless Standards in the U.S.," Personal Indoor and Mobile Radio Communications, Taipei, Taiwan, October 1996.

CHAPTER 5

MOBILE DATA COMMUNICATIONS

5.1 INTRODUCTION

Along with the explosive growth in mobile telephony, demand for mobile data communication services has been growing at a very rapid rate. One of the major drivers for the growth in mobile data communications is the increasing use of portable computers to access Internet and intranet services—from any location and at any time. In many cases, there is also a requirement for uninterrupted service if the location of the terminal changes during the session. The scope of mobile data communication services and applications is very wide, and an illustrative list of these applications would include the following:

- paging and short message services
- electronic mail and Internet/intranet access
- database access and file transfer
- fleet management and dispatch (trucks, taxis, packages delivery systems)
- inventory control (stores, warehouses)
- credit card authorization (from remote or mobile locations)
- field services
- intelligent highways (e.g., automatic toll collection)
- data inputs (e.g., hospital patient charts, auto rental returns)

To support the ever-expanding range of mobile data communication services and applications and their growing demand, either stand-alone (based mostly on proprietary specifications) mobile data networks are being implemented or

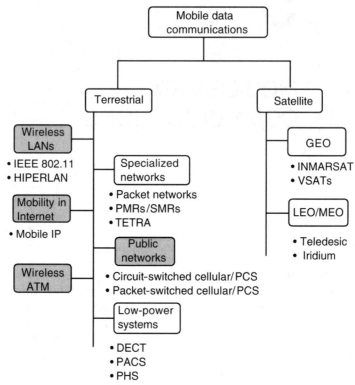

Figure 5.1 Range of systems/networks/standards for Mobile Data Communications: shaded
boxes, primary focus for this chapter.

the capabilities of the existing public mobile (or fixed) communication
systems are being enhanced. Figure 5.1 summarizes the range of systems/net-
works/standards that address one or more aspects of mobile data communications.

5.2 SPECIALIZED PACKET AND MOBILE
RADIO NETWORKS

A number of packet radio networks have been implemented to provide one-
way or two-way messaging services: these are public networks utilizing proprietary technologies and implementations. The only exception comprises the
networks based on the TETRA (terrestrial trunked radio) standard developed
by ETSI. Examples of such one-way (paging) networks in the United States
include EMBARC, MOBILECOM, and SkyTel. Similarly, ARDIS and RAM

rchitecture for GSM in which the mobile is provided dat.

GSM data services architecture of Figure 5.3, the int
(IWF) located at the MSC provides all air interface protoc
unctions that are specific to circuit-switched data services, as
ssary protocol conversion and interworking functions for inter.
landline network. The key functions provided by the IWF are
(RA), the Radio Link Protocol (RLP), which implements the A
adio link, and voiceband modems for communication with the landlin
e role of the transcoder and rate adaption unit (TRAU) is to provide
scoding and rate adaption for the bit stream at the air interface to a
signal utilized internally by the MSC. Sections 5.3.2.1 and 5.3.2.2 offer
scriptions of the rate adaption and error correction aspects.

2.1   Rate Adaption.   The bit rate associated with each interface,
determined by the channel structure, generally is quite different from
at which the end systems communicate. The rate associated with an in-
is always lower than the bit rate of the interconnecting channels. Hence
rate needs to be adapted to align with that of the connecting channels
stage. Figure 5.4 gives an overview of rate adaption stages involved in
nd GSM for supporting data services.

.3.2   Error Control.   Transmission over mobile radio links is subject
ificantly greater interference and delay than fixed network links. Raw
r rates in the range of one per thousand are not uncommon on mobile
inks, making error correction protocols essential for data transmission.
rd error correction methods, which are designed to mitigate bit error
n the one-per-million range encountered in fixed network links, may
unacceptable residual errors. Further, forward error correction (FEC)
terleaving can take care of such impairments as channel fading but will
able to counter the effects of handoffs or channel interruptions.
hus, depending on the required error performance, additional automatic
request (ARQ) protocols may be required. The ARQ protocol imple-
tion in GSM is called Radio Link Protocol (RLP), which is based on
evel data link control (HDLC) and generally resides in the IWF at the
However, use of RLP implies that the network needs to know the ap-
ion and protocols used by the mobile data terminal, and the resulting ser-
called *nontransparent (NT)*, as opposed to *transparent (T)* service when
the FEC and interleaving are applied on the terminal side of the air in-
e.
ransparent service is characterized by fixed transit delay and fixed
ghput but variable bit error rate (BER). Nontransparent service, on the
hand, provides a predetermined (controlled) bit error rate but variable
and throughput. In the North American digital cellular specifications the

Data Mobile represent two-way packet radio networks. Whereas the ARDIS system is based on technology developed by Motorola, RAM Data Mobile uses the MOBITEX technology developed by Swedish Telecom (now Telia Mobitel). There are a number of similar two-way packet radio networks in Europe and other countries based on the MOBITEX technology.

Specialized mobile radio (SMR), or private mobile radio (PMR), is a trunked radio technology originally developed for voice communication, but being increasingly used to support mobile data services. The networks operated in the United States by RaCoNet, Nextel, and Bell Atlantic Mobile fall in this category. The specialized packet radio networks as well as the SMR/PMR systems all essentially deploy proprietary technologies—though there is a MOBITEX Operators Association (MOA), which oversees the specifications for the MOBITEX-based networks, and coordinates software and hardware developments and their evolution.

In Europe, the need for a standardized PMR system was recognized to ensure that mobile data systems for police dispatch and emergency services would be based on a pan-European standard and thus could easily interoperate across national boundaries. This led to the development of the TETRA standard by ETSI. This digital TDMA standard uses 25 kHz carrier channels, with each channel supporting four time slots. It can support user bit rates of 2.4 and 4.8 kb/s, as well as 7.2 kb/s (unprotected). The standard supports dispatch services as well as circuit-switched (voice + data) and packet-switched data services. While the voice + data service allows the use of a mixture of time slots in a 25 kHz carrier, the packet data service is based on the use of the full 25 kHz bandwidth channel for data. The TETRA system has already undergone numerous technical trials and TETRA systems are starting to become operational. Although conceived as a private mobile radio technology for public safety applications (police, ambulance dispatch), TETRA is also seen as a public access mobile radio (PAMR), allowing a TETRA operator to offer shared access to users who cannot afford to implement their own TETRA networks.

5.3   CIRCUIT-SWITCHED DATA SERVICES ON CELLULAR NETWORKS

5.3.1   Circuit-Switched Data on Analog Cellular Networks

Analog cellular networks were primarily designed and optimized to provide telephony services, and low bit rate data services can be supported by deploying suitable modems in the connection. Among the numerous characteristics of radio channels and analog cellular systems that significantly affect their data transmission properties are the following:

- narrow bandwidth of a cellular channel (300–3000 Hz)
- potential for signal loss due to RF fading
- potential for interruption due to handoffs
- use of companding and preemphasis devices

Ordinary landline modems operating at 300 or 1200 b/s can provide adequate data transmission performance in analog cellular environments, except when the channel becomes unavailable during a severe fade or during a handoff. Better data transmission speeds and performance can be achieved through the use of specially designed *cellular modems*. These devices include such additional functions as a smart RJ-11 interface for the cellular phone (for automatic dialing and carrier establishment) and forward error correction, which can maintain connection during short carrier interruptions. The most common forward error correction protocol used for this purpose is MNP 10, which can dynamically vary the frame size and modem data rates to match them to the channel quality. However, even with the use of specially designed cellular or wireless modems, the operational data rates over analog channels remain in the range of 2.4 to 4.8 kb/s—far lower than those of wireline dial-up channels, where rates of 28.8 kb/s are quite common. Figure 5.2 illustrates a typical arrangement for support of dial-up data connections in an analog cellular system.

The *modem pool* (some times also referred as modem converter), which is accessed by the cellular modem user by dialing an access code, facilitates communication between the cellular modem and the landline modem over the cellular network. It provides specialized network and modem-signaling software to connect the cellular network directly to the PSTN for data connections. It can boost the available data speeds up to 9.6 kb/s (given favorable channel

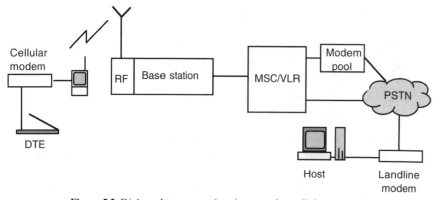

Figure 5.2 Dial-up data connections in an analog cellular network.

...nditions) by providing more robust error corr... ...witch data calls to channels with better transmis... ...ures associated with modem pooling include sin... ...stage dialing for mobile users, as well as consolida...

5.3.2 Circuit-Switched Data on Digital Ce... Networks (Low Speed)

Digital cellular system specifications provide for... data-related bearer services. For example, the ini... cludes the following categories of data (bearer) ser...

- data circuit duplex—asynchronous
- data circuit duplex—synchronous
- PAD access—circuit, synchronous
- data packet duplex—synchronous
- alternate speech/unrestricted digital data
- speech followed by data

A similar set of services also is part of the sp... American and Japanese digital cellular standards (i... PDC). These data services fall in the category of low... services, and the available bit rates range between 30... section uses the GSM examples to elaborate on the sup... data services in digital cellular systems. Asynchronou... circuit duplex represents a commonly deployed servic... ranging from 300 b/s to 9.6 kb/s are possible. Figure 5.3...

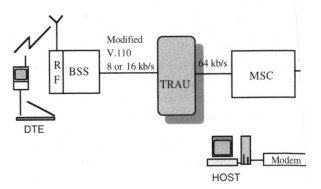

Figure 5.3 Typical circuit-switched data services architectu...

---

service ... the PST... In ... function... nation ... the nec... with th... adaptio... on the ... host. T... the tra... 64 kb/s... brief d...

5.... which ... the rat... terface... the bit... at each... ISDN ...

5.... to sig... bit err... radio ... Forwa... rates ... lead t... and i... not b...

T... repe... men... high-... MSC... plica... vice ... only ... terfa...

thro... othe... dela...

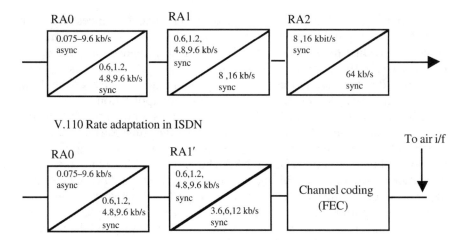

V.110 Rate adaptation in ISDN

Modified V.110 rate adaptation for GSM air interface

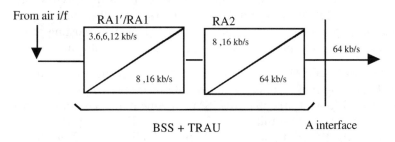

Use of TRAU (transcoder and rate adaptation unit) in GSM

Figure 5.4 Rate adaption stages in ISDN and GSM.

nontransparent and transparent services are referred as type I and type II service, respectively.

5.3.3   High-Speed Circuit-Switched Data in GSM (HSCSD)

The high speed circuit-switched data service in GSM is an extension of the current circuit-switched data services implementation described in the preceding section. The basic approach in HSCSD is to achieve higher user bit rates by using multiple TDMA time slots for a data connection while keeping the GSM physical layer implementation for data services unchanged. The use of multiple time slots within a GSM frame structure of eight time slots per 200 kHz carrier leads to achievable user rates of up to 76.8 kb/s (9.6 × 8 kb/s).

For simplifying the design of the MS and for minimizing changes to the current structure, multiplexed time slots are chosen to be consecutive time slots. Besides providing higher data rates, HSCSD also provides the potential for *pseudoasymmetric* operation, whereby some of the subchannels assigned for the uplink direction are not used, resulting in a larger capacity for the downlink transmissions—an option suitable for Internet applications.

As illustrated in Figure 5.3, most of the functions for data services in GSM are located in the mobile station (terminal adaption function) and the MSC, which provides the necessary interworking functions (rate adaption, error control, and modems) through the IWF. The HSCSD utilizes the same implementation by treating the multiple channels assigned to an HSCSD call as independent subchannels as far as the rest of the GSM network is concerned. The channel splitting and combining functions need to be implemented only at the two ends (i.e., at the MS and the IWF, respectively). The HSCSD architecture is illustrated in Figure 5.5.

Though HSCSD is based on the apparently simple principle of multiplexing up to eight time slots without requiring major changes to the GSM network infrastructure, full advantage of HSCSD in terms of highest bit rates can be achieved only by implementing some additional changes to existing protocols and procedures. Examples of such changes include the following.

BSS–MSC link.   GSM is designed to facilitate interoperation with the ISDN, which has a nominal rate of 64 kb/s or a DS-0 rate for bearer channels. However, the BSS-to-MSC link in most GSM implementations is optimized so that the DS-0 link carries four full-rate traffic channels. This multiplexing function is located at the BSS. With the introduction of HSCSD, a new framing protocol for the BSS–MSC link is required so that HSCSD channels can be carried efficiently and transparently through the MSC to the multiplexing function located at the IWF.

Frequency hopping.   GSM provides for a frequency-hopping option, which requires that each frame be transmitted on a different carrier frequency according to a defined frequency-hopping sequence. To support HSCSD, use

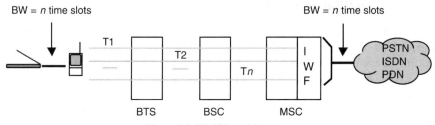

Figure 5.5 HSCSD architecture.

of either multiple synthesizers or an RF duplexer with a single synthesizer with reduced switching interval is required.

Radio Link Protocol (RLP).   Nontransparent data service in GSM is based on the error correction protocol between the MS and the IWF provided by the RLP. The existing RLP specification has been designed for a single, full-rate traffic channel (TCH/F) with maximum user data rate of 9.6 kb/s. An extension to the RLP is required for multilink operation so that multiple frames received from multiple channels for a HSCSD call will appear in correct sequence in the receiving host.

Flexible bearer service.   GSM blocking performance and call-carrying capacity are based on the use of a single time slot or subchannel to carry a call. An HSCSD call that utilizes multiple, consecutive subchannels will therefore encounter increased levels of call blocking, which will increase exponentially with increases in requested data rates. This blocking applies not only at the time of HSCSD call origination but also when such a call needs to be handed off from one cell to another. If the requisite number of subchannels are not available in the receiving cell, the handoff may fail, thereby resulting in the dropping of an existing data connection. However, unlike a voice call, the required or available data rate can be traded off with the delay performance. A single flexible bearer service is therefore defined for HSCSD in terms of a *desired* data rate and a *required* data rate. Thus, depending on the available number of subchannels at the time of call origination or at the time of handoff, the HSCSD call may be assigned any number of subchannels to support data rates between these two limits. The signaling procedures to support this flexible bearer service need to be specified.

Many of the complexities can be avoided and system design can be simplified if the HSCSD implementation is restricted to lower data rates (e.g., 28.8 kb/s) where multiplexing of only two or three subchannels is sufficient. Though the foregoing description for higher rate circuit-switched data services focused on the HSCSD service provided in the GSM phase 2+ specifications, similar services are also foreseen for other digital cellular systems, such as IS-136 in the United States.

5.4 PACKET-SWITCHED DATA SERVICES ON CELLULAR NETWORKS

5.4.1 Packet Data in Analog Cellular Networks—CDPD (Cellular Digital Packet Data)

5.4.1.1 CDPD Network Architecture.   Cellular digital packet data (CDPD) is perhaps the best-known packet-switched service designed for operation with an analog cellular system, specifically, AMPS. Though it can be implemented as a stand-alone service, it is almost always implemented as an overlay capability over an existing AMPS system, where it utilizes unused bandwidth in the system to transmit packets in a connectionless mode. The CDPD system specification was developed in early 1990s by the CDPD Forum initiated by a number of U.S. cellular service providers. CDPD services have been in operation since 1994. CDPD was designed to maximize the use of existing functions and capabilities in AMPS systems, and it shares such AMPS components as RF circuits and antennas, power sources, and location registers. The network architecture for the CDPD system is illustrated in Figure 5.6.

MOBILE END SYSTEM (M-ES).   ME-S represents the subscriber's mobile data terminal, which the CDPD network subscribers use to gain access to wireless communications. These may take different forms ranging from a handheld PDA to a laptop with a PCMCIA card (personal computer memory card international association) to an appropriate point-of-sale terminal. An M-ES may be physically mobile or stationary, but the CDPD network always considers it to be potentially mobile. The CDPD network continuously tracks the location of an M-ES and thereby ensures that datagrams addressed to it continue to be delivered even if its physical location changes. An M-ES may support either full-duplex or half-duplex operation, and it can support power levels ranging from 0.6 to 3.0 W with capability to dynamically adjust the power level. A *sleep mode* option allows an M-ES to power down when not transmitting data and periodically power up to check for buffered incoming data.

MOBILE DATA BASE STATION (MDBS).   The MDBS essentially operates as a mobile data link relay that acts on behalf of an MD-IS and therefore needs to support the CDPD MAC and data link layer procedures across the radio interface, as well as the protocols across its landline connection to

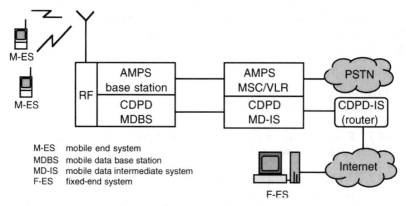

Figure 5.6 CDPD network architecture.

MD-IS. MDBS is responsible for all radio resource management or control functions which, in the case of CDPD, include radio channel allocation, inter-operation of channels between CDPD and voice calls, and tracking of busy/idle status of reverse or uplink channels (for potential allocation to CDPD packets). The MDBS is most often colocated with the AMPS base station and shares the AMPS antenna and RF plans.

MOBILE DATA INTERMEDIATE SYSTEM (MD-IS).   In the case of typical cellular systems designed for voice services, the MSC/VLR supports such functions as mobility management (location tracking, registration, authentication, encryption), call and connection control (circuit switching), roaming support (interfacing with a visited cellular network), and billing and accounting. In the case of CDPD, the MD-IS is primarily responsible for the mobility management, roaming support, and billing and accounting functions. Billing and accounting is perhaps the most critical and complex function in the CDPD's packet environment, where tariffs are based on amount of data (packets) delivered. The equivalent of the switching function in a packet environment is *routing,* and this, in the case of CDPD, is provided by the CDPD-IS or simply an IS, which in most implementations is an *off-the-shelf* router suitable for routing IP or CLNS datagrams. MD-IS (with its associated IS or router) may perform the so-called mobile home function (MHF) or mobile serving function (MSF). The mobile home functions are equivalent to the functions performed by the MSC/HLR in a cellular voice network, whereas the mobile serving functions are equivalent to those of an MSC/VLR.

5.4.1.2   CDPD Reverse and Forward Channel Formats.   All communication in CDPD takes place via an MDBS, and direct communication between two M-ES within a cell is not allowed. Multiple M-ES share the reverse (M-ES-to-MDBS) channel to communicate with an MDBS, an operation that requires well-defined protocols for channel sharing and contention resolution on the reverse channel. Further, a robust forward error correction (FEC) method is required on both the forward and reverse channels to compensate for the error-prone radio environment. The FEC chosen for CDPD is the Reed–Solomon (63,47) encoding, with 47 data symbols (6 bits/symbol) followed by 16 parity symbols. The format of the CDPD reverse channel is shown in Figure 5.7.

Information to be transmitted over the radio channel is formatted as variable-length frames, which are delimited by unique frame flag sequences. The frames are then transmitted in the form of a *burst* containing an integral number of Reed–Solomon FEC blocks interleaved with necessary control and synchronization bits. In the reverse channel, a 38-bit ramp-up or dotting sequence and a 22-bit sync sequence precede the transmission of R-S blocks. Each R-S block is interleaved with continuity indicator bits (total of 7), which indicate the termination (or otherwise) of the transmission. After the last

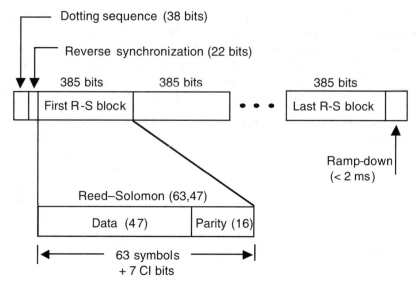

Figure 5.7 CDPD reverse channel signaling format: CI, continuity indicator.

block of information, there is transmission ramp-down period that may last up to 2 ms.

In the case of the forward channel (MDBS-to-M-ES), data transmission consists of a continuous (and contention-free) stream of R-S blocks, which are interleaved with control flags. In the 6-bit control flags, which occur every 10 microslots (symbols) in the R-S block, the first 5 bits are used for busy/idle indication and the sixth bit for block decode status. The busy/idle bits indicate whether the reverse channel is idle or busy, and this information is used by the M-ES to time the transmissions. The decode status flag provides the indication of success or failure in decoding the block just received from the M-ES. Failed decode indication is perceived as a collision by the M-ES (even though the failure may be due to channel impairments), and it then invokes the contention resolution protocol, which in CDPD is referred as slotted, nonpersistent digital sense multiple access with collision detection (DSMA/CD).

5.4.1.3 CDPD Protocol Architecture. The CDPD protocol architecture has been designed to ensure efficient use of radio link resources and to maximize error-free data transmission in a typically hostile radio environment. Figure 5.8 presents an example implementation of the CDPD protocol stack. The subsections that follow discuss some of the components of this system.

NETWORK LAYER. CDPD supports both the Internet Protocol (IP) and the OSI Connectionless Network Protocol (CLNP). An M-ES is assigned a

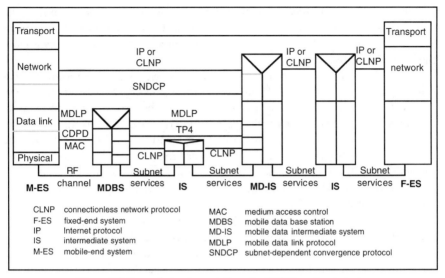

Figure 5.8 Example implementation of a CDPD protocol stack.

unique network address by the CDPD service provider for packet delivery and billing. The address is associated with an MD-IS performing the mobile home functions (MHF) that can support roaming.

SUBNETWORK-DEPENDENT CONVERGENCE PROTOCOL (SNDCP). To conserve radio bandwidth, CDPD provides for compression of the unusually long network layer packet headers used in TCP/IP and CLNP. The SNDCP layer in CDPD can compress the 40-octet TCP/IP headers to 3 octets, and the 57-octet CLNP headers to 1-octet compressed headers. Further, CDPD provides an option of using V.42bis compression for the payload data. The compressed network layer packets are then segmented into 128-byte protocol data units (PDU), which are encrypted by means of the Diffie–Hellman public key algorithm for transfer to MD-IS.

MOBILE DATA LINK PROTOCOL (MDLP).   The MDLP is based on the ISDN LAPD procedures with appropriate modifications for conserving radio bandwidth. For example, MDLP uses selective reject (SREJ) procedures for retransmission of lost packets, and it supports multicast and broadcast addressing for the unacknowledged data service. To facilitate efficient management of handoffs, MD-IS also terminates the MDLP.

CDPD MEDIUM ACCESS LAYER (CDPD MAC).   This layer uses DSMA/CD (digital sense multiple access with collision detection) procedures to control access to the radio channels. In the DSMA/CD mode, the M-ES wanting to transmit a message senses the busy/idle flag in the forward channel and, if the channel is busy, it waits for a random period before sensing the channel again. This is referred as the *defer mode*. On the other hand, if it senses that the channel is idle, it initiates transmission within an 8-bit interval from the end of the busy/idle flag.

The M-ES continues transmission until it senses a decode failure indication on the forward channel. It then ceases transmission and enters the *back-off mode* in which it waits for a random interval before sensing the channel again for busy/idle status. Thus the *defer mode* is invoked when the channel is sensed busy and the *backoff mode* when a decode failure occurs. While the random waiting interval in the defer mode is based on a uniform distribution, the waiting interval in the backoff mode is based on a uniform distribution but is followed by an exponential algorithm in subsequent retransmissions due to multiple failures.

PHYSICAL LAYER.   Since the CDPD operates as an extension to or overlay on the analog AMPS system, it uses the AMPS channels, which are 30 kHz wide. The bit stream over the physical interface is modulated using GMSK and has a data rate of 19.2 kb/s. As described in the preceding section, the data are protected against transmission errors over the radio link using Reed–Solomon (63,47) encoding.

5.4.1.4   Channel Assignment and Hopping.   There are a variety of ways in which the radio channels may be configured and assigned in a CDPD implementation that is overlaid on an existing AMPS system. Some of the options are as follows:

Fixed assignment.   In this case a fixed number of channels from the AMPS pool are permanently assigned to the CDPD network and there is no sharing of channels between the voice and data parts of the implementation. Though such an assignment greatly simplifies the CDPD system design (and thereby leads to cost reductions), it defeats the underlying purpose of CDPD, which is to share the unused capacity of the AMPS system in real time.

Partial sharing.   In this case again a certain number of channels are designated for CDPD use, but these channels are shared between the voice traffic of the AMPS system and the data traffic in the CDPD system. In this assignment, the voice calls first try the channels exclusively assigned to the AMPS system. If they are all busy, the voice traffic overflows to the channels shared between the AMPS and the CDPD systems. Thus, this assignment scheme provides a certain degree of protection for the data traffic.

Full sharing.   In this assignment scheme, the complete pool of AMPS channels is available for sharing between the voice and data traffic in real time. Since, however, the assignment is subject to preemptive priority in favor of voice calls, the data transmission on a channel is suspended (until another idle channel is found) if the channel is claimed by a voice call.

Both full and partial sharing schemes require that the system implement some type of sharing protocol between the voice part and the data part of the network. To this end, a channel status protocol (CSP) may be implemented at the MDBS, to allow message exchange with the AMPS system to update the shared channel status. Alternatively, the MDBS may deploy a *sniffer* that periodically scans the shared channels to update the busy/idle status for the channels that can then be broadcast to the M-ESs over the forward channel. In the latter case, the operation is transparent to the AMPS system.

CHANNEL HOPPING.   In the shared channel implementation, when a voice call arrives at the AMPS base station, the BS selects an idle channel to serve the call. However, if the channel is busy serving a CDPD data call, the MDBS is required to give up the channel within 40 ms. The MDBS then attempts to reestablish the channel stream associated with the suspended call on another free RF channel. This type of channel switching in CDPD is referred as *forced or emergency channel hopping.* If a free channel is not available to accommodate the preempted data call, the associated channel stream enters the *blackout* state.

In an AMPS system, if a channel is perceived to be subject to undue interference, it may be sealed or temporarily made unusable. This can also occur when the AMPS system notices interference on the channel due to CDPD activity, and in this case the CDPD system will effectively steal the channel from the AMPS systems. To avoid such channel stealing, the MDBS may execute a *timed hopping* or *planned hopping,* periodically switching the channel stream for a CDPD data flow from one channel to another. A number of timers are defined and implemented for this purpose. For example, a *dwell timer* and a *layoff* timer are defined—the former specifies the maximum interval for which an RF channel can be used for a CDPD data stream, and the latter specifies the duration (generally 40 ms) before a channel released by a *timed hop* can be reused by a CDPD data stream.

5.4.2 Packet Data in Digital Cellular

The most common packet service that is supported on digital cellular networks is the so-called short message service (SMS). The service is available in most digital cellular networks (e.g., GSM, IS-136, IS-95, PDC). It is a store-and-forward messaging service that provides delivery of short messages originated by digital cellular phones and delivered to digital cellular phones with small screens. However, more comprehensive and powerful packet services are now being defined. Examples of these emerging packet services include GPRS (general packet radio service) in GSM and IS-136, PDC-P in the Japanese PDC system, and EDGE (enhanced data rates for GSM evolution).

5.4.2.1 Short Message Service (SMS).   The data services supported in various digital cellular systems represent extensions of data services provided to subscribers in a fixed network but not really tailored for small, handheld mobile terminals. These data services generally require significantly larger terminals or terminal attachments to a handheld mobile terminal. SMS, however, allows transmission of short (up to 140 octets), point-to-point or broadcast messages, which can be displayed on the screen built into the cellular handset. Some of the short message categories that are available in digital cellular systems are summarized as follows.

Short message service—mobile terminated/point-to-point (SMS-MT/PP) is intended for the reception of short messages from within or outside the GSM environment. The way the message is sent from the original sender to the network (for delivery to the intended user) may depend on the cellular network operator. The incoming message is stored at the SMS service center (SMS-SC) before the network tries to deliver it to the user. Options similar to voice messaging systems may include message-waiting indication, message delivery acknowledgment, and message storage at the terminal.

Short message service—mobile originated/point-to-point (SMS-MO/PP) enables a cellular network user to send a message to another cellular network user; in the future this service will be extended to other parties, such as an electronic mailbox or gateway devices, which, for example, could translate the message and deliver it as a fax. The availability of both mobile-terminating and mobile-originating SMS services (unlike the paging case) allows digital cellular network users to engage in (non-real-time) dialog.

Cell broadcast short message service (SMS-CB) is another short message service that allows short messages of a general nature (weather reports, road conditions, etc.) to be broadcast at regular intervals to all subscribers within a specified geographic area.

The SMS implementation has two significant features:

- SMS is a store-and-forward (packet mode) service, and the SMS service center has been defined to act as a message store, which provides the necessary message-forwarding capabilities as well as interworking with the various applications and services within the fixed network.
- For message transfer between relevant network entities, control and signaling channels (instead of normal traffic channels) are generally used for data transmission.

For example, Figure 5.9 uses a simplified SMS architecture to indicate the steps involved in message delivery for mobile-terminated SMS in GSM.

An enhancement to the SMS in GSM is also underway as part of phase 2+ GSM specification, which will allow concatenation of the current 140-octet SMS messages to form longer messages by means of a continuity flag. The con-

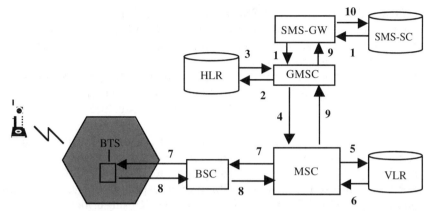

1. The MSISDN stored in the SMS-SC is used to access the GMSC of the mobile's HOME network.
2. The GMSC requests the HLR for the address of the serving MSC for the MS.
3. The HLR returns the address of the serving MSC/VLR.
4. The GMSC routes the Short Message (SM) to the serving MSC.
5. The serving MSC requests the paging area from the VLR.
6. The VLR returns the paging area for the MS.
7. The MSC pages the MS and establishes a signaling channel over the air interface and delivers the SM.
8. The MS sends the acknowledgement for the SM to the MSC.
9. The MSC reports the outcome to SMS-GW.
10. The SMS-GW forwards the report to SMS-SC.

Figure 5.9 Steps for (successful) short message delivery.

catenated message will be sent over the signaling channel (SDCCH) as multiple packets (each containing 140 octets). Enhanced SMS is more efficient but is restricted to simplex or unidirectional information transfer.

5.4.2.2 General Packet Radio Service (GPRS) in GSM.   The existing GSM circuit-switched data services and the high speed circuit-switched data (HSCSD) services described in Section 5.3 provide reliable connectivity and high throughput for data services. However, significant portions of today's mobile data communications are based on bursty Internet applications like e-mail and World Wide Web browsing, for which use of a circuit-switched connection is wasteful and expensive. To capture the market potential of such bursty data applications, GSM Release '97 introduced the general packet radio service.

GPRS maintains the core GSM radio access technology and provides packet data services by introducing two new network elements called GPRS support node (GSN) and gateway GPRS support node (GGSN). In addition, the GPRS register, which may be integrated with the GSM HLR, maintains the GPRS subscriber data and routing information. GPRS will operate at transmission data rates from 14.4 to 115.2 kb/s by using from one up to eight time slots in GSM TDMA structure. The network architecture for GPRS is illustrated in Figure 5.10.

The architecture attempts to ensure that few or no hardware modifications are required for the existing GSM network elements and that same transmission links between the BTS and the BS can be utilized. Further, GPRS will utilize the existing GSM authentication and privacy procedures, and the GPRS register, which contains the GPRS subscriber data and routing information, can be integrated with the HLR. Besides the GPRS register, two new nodes are defined for GPRS. The serving GPRS support node (SGSN) is responsible for communication between the mobile station (MS) and the GPRS network. The primary functions of the SGSN are to detect and register new GPRS mobile stations in its serving area, to send/receive data packets to/from the GPRS MS, and to track the location of the MS within its service area. The GGSN acts as the GPRS gateway to the external networks and provides such interworking functions as translation of data

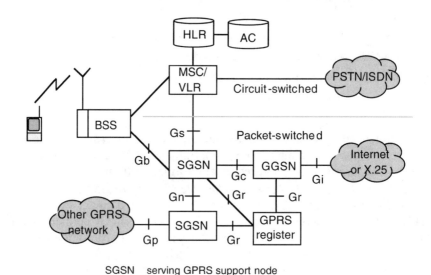

SGSN   serving GPRS support node
GGSN   gateway GPRS support node

Figure 5.10 GPRS network architecture.

formats, protocols conversions, and address translation for routing of outgoing and incoming packets.

The protocol model for GPRS has been designed to facilitate seamless access to external data networks like the Internet (TCP/IP) and PSPDNs (X.25/X.75). The protocol architecture model for GPRS is shown in Figure 5.11.

The transport and network layers are optimized for support and interoperation with different external data networks like the Internet and X.25. The SNDCP is a convergence layer that provides the functionality to map different network protocols onto the logical link control (LLC). This functionality includes multiplexing of packets from different protocols, header compression, data compression (e.g., V.42bis), data encryption, and segmentation of packets to match the LLC packet size.

Communication across the radio link (MS-to-BSS and vice versa), involves the physical layer, and the RLC and MAC components of the data link layer as indicated in Figure 5.11. The logical link control, the higher sublayer of the data link layer, is primarily used for establishing a logical link between the MS and the SGSN and can support both an unacknowledged mode and the acknowledged mode—the latter providing packet retransmissions and flow control for error-free delivery. The LLC protocol is based on LAPD, which is used in GSM and can support point-to-multipoint (PTM) transmissions.

The Radio Link protocol (also used in other GSM data environments), operates in an acknowledged mode with a sliding window flow control and se-

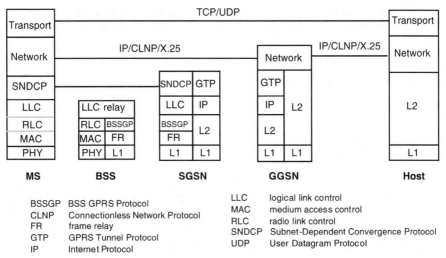

Figure 5.11 GPRS protocol architecture model.

lective automatic repeat request (ARQ) error correction to provide an error-free link between the MS and the BSS. The maximum size of an LLC packet is 1600 bytes, and the RLC segments them into smaller RLC blocks whose size depends on the encoding scheme in use.

The medium access control (MAC) layer in the GPRS is designed to facilitate packet data transmission across the radio interface. It is based on the slotted ALOHA protocol. In addition to controlling access to the radio link, the MAC layer performs contention resolution between multiple MSs attempting channel access, arbitrates between multiple service requests from different data terminals, and allocates the radio channel to individual data terminals on request. The RLC/MAC layer combination in GPRS ensures flexible allocation and utilization of radio resources in the multilink GPRS environment. Flexibility is achieved by defining a packet data traffic channel (PDTCH) multiplexed onto a physical data channel. Up to eight PDTCH channels can share one time slot in the TDMA frame, so that eight MSs can share a single TDMA time slot for packet transmissions. Further, the multislot capability in GPRS allows an MS to transmit data on multiple time slots (up to eight) within a TDMA frame. Besides the PDTCH, a number of additional logical channels are defined in GPRS to facilitate such functions as broadcast, random access, paging, access grant, multicast, and associated control.

The physical layer is also adapted for GPRS requirements. The physical channel can be encoded by means of one of the four available coding methods, the choice being based on a trade-off between bit error rate and throughput across the channel. Interleaving, which is implemented across one RLC block (each block consisting of four bursts), results in reduced interleaving delay compared to regular GSM operation in the circuit-switched mode.

GPRS has service and network features that makes it an attractive mobile data communication service. Some of the key service features are as follows:

- bandwidth on demand for point-to-point transmission
- negotiated quality of service (QOS)
- point-to-point (PTP) and point-to-multipoint (PTM) service
- multicast and group call services
- value-added services like broadcast information services (e.g., traffic reports, stock prices)
- design for easy Internet access and Web browsing

5.4.2.3   Packet Data Service in the PDC System (PDC-P).   A packet data service very similar to GPRS in GSM has been specified for the Japanese Personal Digital Cellular system, and the network is referred as PDC-P. It supports the Internet protocol TCP/IP and is intended to provide seamless interconnection to the Internet. It utilizes newly defined physical channel structure and link control procedures to support packet transport at high data rates. The key criteria used in the design of the PDC system include the following:

- high packet data rates and efficient radio resource utilization
- transparency and independence from user protocols and applications
- support of multiple network providers and roaming
- compatibility with and maximum reuse of PDC capabilities
- support of handoffs
- maximization of battery life
- high level of data and connection security

A simplified view of the PDC-P network architecture is shown in Figure 5.12.

The packet processing module (PPM) performs functions similar to the SGSN in GPRS. It communicates with the PDC-P mobile data terminal (via the base station) for transmission and reception of user and control data packets. It utilizes the signaling connection to the visited mobile switching center (VMSC) for paging the data terminals for incoming messages. The primary functions provided by the packet gateway module (PGW) are connection and interworking with external networks, interacting with the location registers (LR) to support mobility management, and routing of data packets within the PDC-P network. The location register in PDC-P, which stores location infor-

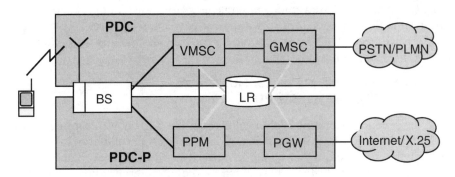

PPM   packet processing module
PGW   packet gateway module
LR   location register

Figure 5.12 Network architecture for the PDC-P packet data network.

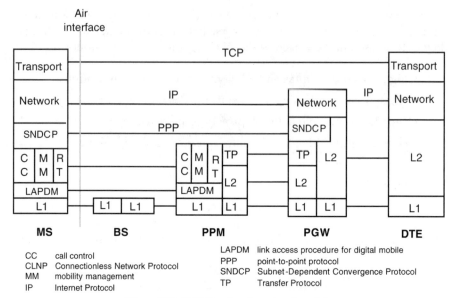

Figure 5.13 PDC-P network protocol model.

mation about each mobile data terminal, as well as subscribers' service profiles, is shared with the voice part of the PDC system.

PDC-P utilizes the TDMA structure of PDC, where a carrier frequency is divided into three duplex time slots. By utilizing all the three time slots on a carrier, PDC-P can deliver a maximum transmission rate of 42 kb/s and a user data rate of 28.8 kb/s. PDC-P utilizes a newly defined user packet channel (UPCH), which is shared among multiple users for transfer of packet data. The channel is BCH-encoded for forward error correction and uses a 16-bit CRC for error detection. The protocol model for PDC-P, which is very similar to that for GPRS, is shown in Figure 5.13.

5.4.3   Evolution of Cellular Mobile Data Capabilities: The EDGE Concept

All the second-generation digital cellular system designs were optimized for telephone traffic. The initial support of data services on cellular networks was essentially restricted to dial-up modem-based services. The user data rates for these services were further constrained by the intereference-prone nature of the radio environment. Attempts to support such services as HSCSD and GPRS essentially represent add-on capabilities to the basic voice-optimized cellular networks that nevertheless maintain the essential characteristics of the radio access technology.

However, it is becoming increasingly evident that if the second-generation cellular mobile systems wish to enhance their data-handling capabilities to support the emerging high speed data and multimedia market, the radio access part must be modified. Initial efforts in this direction have been under way for some time in Europe, and the resulting EDGE (enhanced data rates for GSM evolution) proposal is being developed in ETSI SMG. Support of EDGE in GSM is considered to be a key point in its evolution to the third-generation mobile system (UMTS/IMT-2000). The EDGE concept applies to both circuit mode and packet mode data and is sufficiently generic for application to other digital cellular systems. For example, it has been adopted by the TIA in the United States for evolving the data capabilities of the IS-136 TDMA system toward third-generation capabilities.

The EDGE concept proposes to use the 200 kHz bandwidth in GSM with one or more high level modulation schemes and a range of efficient encoding methods that can be combined to provide higher bit rates across the radio interface. The development of the EDGE proposal toward a specification is under way in ETSI SMG 2, and various high level modulation schemes and encoding methods are under study and analysis. A feasibility study was completed at the end of 1997. A number of proposals have followed from this study, including the two modulation schemes: offset QPSK (OQPSK) and offset 16 QAM (O16QAM). Performance studies have indicated that these modulation schemes provide the best trade-off between spectral efficiency and implementation complexity. Similarly, a number of multilevel encoding schemes that provide good performance for data transmission at moderate complexity are also being studied.

The RLP block structure for packet retransmissions in EDGE is identical to structures found in the GPRS specification. A data packet to be transmitted is segmented into a number of RLP blocks, each of 20 ms duration, with information content in each block varying from 224 bits to 1304 bits depending on the modulation/encoding combination used. The RLP blocks carry a 16-bit CRC field, which is checked at the receive end for retransmission decisions. EDGE will also support a *link adaptation* mechanism, which selects the best possible combination (in terms of maximizing link data transfer rate) of modulation and encoding scheme based on the time-varying link quality.

The EDGE specification whose scope includes both packet-switched and circuit-switched data was to have been finalized in the 1999 time frame. Studies for evolution of IS-136 TDMA cellular system toward third-generation mobile system capabilities, in line with the EDGE principles, are also under way in the United States through TIA committee TR45.

5.5   DATA OVER LOW POWER WIRELESS OR
CORDLESS TELECOMMUNICATION NETWORKS

Cordless systems like CT2, DECT, PACS, and PHS, described in Chapter 4, were primarily designed for support of voice telephony service over a limited coverage area and for low mobility applications. However, with the increasing demands for data communications in the office and indoor environments, the data capabilities of these cordless systems are coming into better focus. For example, the allocation in the PCS band of 1920–1930 MHz for unlicensed isochronous applications and 1910–1920 MHz for asynchronous LAN applications in the United States has spurred interest in the data capabilities of low power wireless systems. Sections 5.5.1 to 5.5.4 briefly describe some of the data services capabilities for the DECT, PACS, and PHS systems.

5.5.1   Data Services in DECT (Digital Enhanced
Cordless Telecommunications)

DECT system has a TDMA/TDD frame structure with 24 slots that are equally allocated for downlink and uplink operation. DECT specifies both simplex (half-slot) and duplex (full-slot) operation. Each slot has a protected and an unprotected format in which error protection can be supported within the protected format and utilized for error-sensitive services like data. DECT further provides the capability to combine several time slots to provide a high bit rate connection. Within the multislot or *multibearer connection* (as it is known in DECT), both symmetric and asymmetric operations are permitted. With the growing demand for the support of higher data rates, DECT has enhanced its data rate capabilities. The higher data rates are achieved by utilizing multilevel modulation in DECT. The basic modulation scheme in DECT is a two-level GFSK (Gaussian filtered frequency shift keying), which is supplemented with four- and eight-level modulation schemes for the data field leading to data rates as high as 2.88 Mb/s per carrier when eight-level modulation is deployed.

The radio channel characteristics of DECT are subject to large variations in signal quality due to multipath fading and resulting error bursts. Studies have indicated that FEC techniques generally are not effective in this severe error-prone environment, and ARQ is a better choice to maintain data integrity. However, ARQ procedures lead to delays that are proportional to the channel error rate, thereby limiting the use of the automatic mode to asynchronous services. DECT can support synchronous services with variable throughput but low delay.

The flexibility in DECT time slot assignments supports a wide range of data, voice, and integrated data/voice services. These services and applications are configured by *mixing and matching* the combination of slots (full, half, or double), number of time slots, and slot formats (protected or unprotected). The range and scope of DECT data services are defined in the so-called DECT *data service profiles,* specified as part of the DECT specification. These data service profiles are categorized as types A–F. Data services addressed by the profiles are exemplified as follows:

Type A	low speed frame relay up to data rates of 24 kb/s
Type B	high performance frame relay with data rates to 422 kb/s
Type C	nontransparent data streams requiring link access protocol (LAP)
Type D	transparent and isochronous connections
Type E	short message and apaging services
Type F	support of teleservices like fax and applications like e-mail and file transfer

In a structure newly developed for DECT data profile standards, called the DECT packet radio service (DPRS), the existing profiles have been combined into a single standard. The DPRS standard classifies the DECT data services into three basic services identified by types A, B, and C and by two mobility classes, 1 and 2. Under this classification, service type A is optimized for low power and simplicity and service type B for higher speeds and throughput. Service type A and B offer a generic frame relay service at the user plane. While service type A supports only symmetrical single-bearer connections, service type B provides media access control (MAC) support for multibearer and asymmetric connections. Both these service types are geared mainly for connectionless applications or for connection-oriented applications that do not require DECT to provide the data link control layer.

Service type C adds the link access protocol at the user plane (LAP-U) functionality to the two basic service types represented by A and B. Therefore service type-C represents a superset of the two basic service types. It offers a data link control layer at the user plane in addition to the frame relay service provided by type A and B services. It provides flow control, ARQ, and inband user signaling. The type C service is primarily intended for connection-oriented applications that require data link control.

Mobility classes 1 and 2 are intended for applications in closed user groups (CUGs) and private and public roaming environments, respectively. Mobility class 1 does not support call setup and takedown procedures, so the virtual circuit is implicitly and permanently present. The service offered by mobility class 1 is analogous to a permanent virtual circuit (PVC) service. Mobile class 2, on the other hand, offers full control plane (C-plane) functionality, including call setup procedures, mobility management, and service negotiation. For this mobility class, the virtual circuit is present only in the context of a call, service parameters can be negotiated at call setup time and may be altered during the call.

5.5.2 Data Services in PACS (Personal Access Communication System) [22]

The PACS radio interface standard supports a variety of data and messaging services, which include

- individual messaging service
- circuit mode data service
- packet mode data service
- interleaved speech/data service

The *individual messaging service* in PACS can support maximum message lengths of 16 mega-bytes. Error-free and efficient delivery of messages is ensured by suitable error control and flow control procedures defined in the PACS specification. Provision for message encryption is included. There are numerous applications for this type of service, including text messaging, e-mail, group 3 fax, and graphics imaging. The service is implemented by modifying the PACS call control messages based on layer 2 acknowledge mode protocol (AMP), which can be used to deliver individual messages efficiently and with minimal complexity.

The *circuit mode data service* in PACS is a nontransparent data service that uses the link access protocol for radio (LAPR) to provide error and flow control, and data encipherment for privacy. The estimated round-trip delay across the radio interface is in the order of a few tens of milliseconds, and this includes the transport delay across RP-RPCU link and the processing times in the RPCU. The round-trip delay compares well with other implementations of nontransparent data services that use radio link protocols (e.g., RLP in GSM). The data throughput for a 32 kb/s channel is better than 28 kb/s under most operating conditions.

The *packet mode data service* in PACS uses the data sense multiple-access (DSMA) protocol over the radio interface for contention resolution, which is a shared contention-based RF packet protocol. On the downlink, the delivery of packets is based on a very efficient scheduling mechanism. The structure of the packet channel in PACS permits efficient and fair sharing of the channel by different types of subscriber unit. Thus, subscriber units that are capable of operating on the basis of a single time slot per TDMA frame can share the channel with subscriber units that can support multislot operation. In the latter case, higher throughput and lower delays can be achieved.

In thc *interleaved speech/data service* both speech and data information can be transmitted over a single 32 kb/s time slot. Data are interleaved with speech bursts during the quiet periods and can be enciphered for greater privacy. Greater handoff reliability is possible because only one new 32 kb/s time slot or channel needs to be set up to the receiving radio port. As in the case of circuit mode data service, the data bursts are delivered across the radio interface by the LAPR protocol.

To provide end-to-end flow of data, an interworking function (IWF) is required to convert digital data on the radio interface to a form suitable for transmission over the intermediate (transit) and the remote (destination) networks. The IWF needs to reside in a network or network entity from which the remote network (and the host) can be reached. In the example architecture for support of data services in PACS shown in Figure 5.14, the intermediate network is assumed to be an ISDN network which uses X.25 on the ISDN B or D channel. In practice, the intermediate network may take different forms, which include a circuit-switched (ISDN) network, a packet-switched (X.25) network, or a local-area network (LAN).

Besides providing the digital data conversion functions mentioned above, the IWF provides a number of other functions that may include the following:

- protocol conversion and adaption to ensure data flow across the radio access, the intermediate, and the remote networks
- handoff management between different radio port control units
- address management and translations
- shielding of the remote network from radio-specific functions (like handoff)

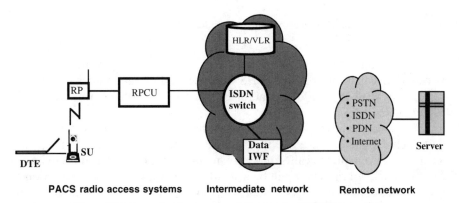

PACS radio access systems   Intermediate network   Remote network

Figure 5.14 Example of network architecture for PACS data services.

5.5.3   Data Services in PHS (Personal Handyphone System)

The range of data capabilities and data rates available in the initial PHS system include group 3 fax at 4.2 to 7.8 kb/s and full duplex modem transmission at 2.4 to 9.6 kb/s through the speech codec. Subsequently these capabilities were enhanced to support 32 kb/s data by direct access to one of the bearer channels. This 32 kb/s data transmission speed will enable faster Internet access, picture transmission at high speed, video transmission and other multimedia applications. A number of new data applications for PHS using small, handheld personal digital assistants (PDAs) are emerging, including the multimedia access system (MMAC) for personal geopositioning via PHS. PHS handsets capable of supporting multimedia applications are starting to emerge in the market, including products that have a wireless optical interface that can be connected to a PDA. PHS handsets with implementation for PCMCIA (Personal Computer Memory Card International Association) are also available and will greatly facilitate connection of computers to the PHS service. The reference configuration for ISDN-like 32 kb/s data services in PHS is shown in Figure 5.15.

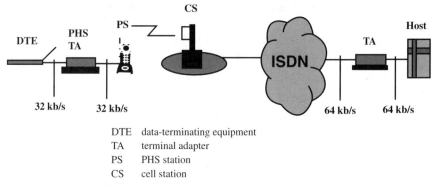

DTE   data-terminating equipment
TA   terminal adapter
PS   PHS station
CS   cell station

Figure 5.15 Reference configuration for PHS 32 kb/s data service.

The primary focus of the current PHS evolution is toward developing packet data capabilities in PHS. PHS with packet data capabilities is generally referred as PHS-PD, and the general architecture of PHS-PD is illustrated in Figure 5.16.

The special-purpose layer 3 protocols needed to manage the wireless circuit, including terminal authentication, IP address assignment, handoff, and sleep mode control, are called mobility management protocols. It is envisaged that the packet data server will be a stand-alone entity because integration of these functions in an ISDN switch will require upgrading of existing ISDN switches, and a standardized interface between the cell station (CS) and the

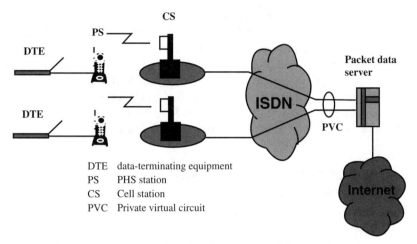

Figure 5.16 Reference architecture for PHS-PD public network service.

public network operator. In the PHS-PD architecture of Figure 5-16, the packet data server provides the mobility management protocol (MMP) for the packet data service. The MMP is used to set up and maintain the wireless part of the TCP/IP connection between the packet data terminal (DTE) and the Internet. The connection is established by means of a radio signal.

The initial implementation of PHS-PD service will utilize the available control channels in PHS radio interface. With ordinary circuit switching, the control slots use only one frame out of 20, with the remaining 19 slots therefore available for transporting packet data. The use of a single-slot channel can provide a data transmission speed of 20 kb/s. For higher data rates it would be necessary to develop procedures for controlling packet transmissions over multiple slots and over communication channels.

5.5.4   Data Services in CT2 (Cordless Telephony 2)

The FDMA-based *CT2 system,* with 32 kb/s duplex channels, also offers asynchronous and synchronous data services ranging from 300 bits/s to 19.2 kb/s. Synchronous mode data services without error correction can operate at 32 kb/s. Whereas asynchronous data services (like LANs) do not have any delay and time constraints, synchronous services (e.g., data interworked with PSTN/ISDN) have time and delay constraints. The indoor environment is prone to long fades, and support of data services in the CT2 frequency band requires that robust error detection/error correction mechanisms be built into the applications.

ETSI, which oversees the CT2 standard in Europe, has proposed the use of Reed–Soloman forward error correction and ARQ for asynchronous services. The ARQ is implemented in the radio link protocol for CT2 called link access protocol for radio (LAPR), which utilizes HDLC frame formats with FEC. For the synchronous mode, the data stream is rate-adapted to one of the standard rates (min = 300 bits, max = 19.2 kb/s). For 32 kb/s the bit stream is not subjected to rate adaption. The frames are transmitted in 1 ms bursts over a 16 ms period. The CT2plus system described in Chapter 4, which is a Canadian enhancement to CT2, also utilizes R-S codes for FEC but uses only 6 bits/symbol (as opposed to 8 bits/symbol in the ETSI CT2 proposal).

5.6 WIRELESS LOCAL-AREA NETWORKS (WIRELESS LANs)

5.6.1 Background

The popularity and the market potential for wireless local-area networks (WLANs) is driven by such applications as inventory control in warehouses and stores, point-of-sale terminals, rental car checkins, and patient record updates in hospitals. These applications take advantage of the mobility associated with WLANs. Further, WLANs find applications in environments where cable installations are either not cost-effective or not practical—for example, at trade shows and conventions, on manufacturing floors, and in heritage buildings. In the interest of enabling all parties to use the available frequency band efficiently and fairly, and to ensure interoperability of terminals and equipment from multiple vendors, wireless LAN standards are emerging in major regions of the world. The two primary standards that have emerged are IEEE 802.11 in the United States and HIPERLAN (High Performance European Radio LAN) in Europe.

Frequencies assigned for wireless LANs are in the unlicensed bands. The U.S. Federal Communications Commission (FCC) has allowed the use of the frequencies in the industrial, scientific and medical (ISM) bands for WLANs using spread spectrum methods. The ISM frequencies are in the 902–928 MHz, 2.4–2.4835 GHz, and 5.725–5.850 GHz bands. IEEE Standard 802.11 applies to the 2.4 GHz band.

5.6.2 IEEE 802.11 Wireless LANs

IEEE Standard 802.11 specifies the physical and MAC layers for operation of WLANs and addresses the direct sequence spread spectrum (DSSS) and frequency-hopping spread spectrum (FHSS) access methods for the radio medium. It also allows for a third option for the infrared medium, which is still under development. The standard provides for data rates of 1 and 2 Mb/s—the latter being optional. It specifies support for asynchronous as well as synchronous data transfers—again the latter being optional. The asynchronous data transfer applies to applications that are not time sensitive (e.g., e-mail and file

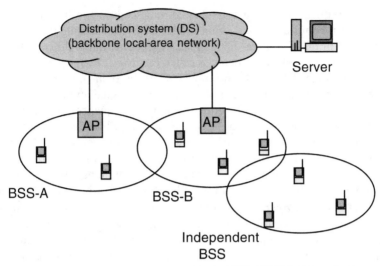

Figure 5.17 Architectures supported in the IEEE 802.11 LAN standard.

transfers). The synchronous mode supports time-bounded applications like video and packetized voice. The two main architectures supported by IEEE 802.11 standard are illustrated in Figure 5.17.

In the independent BSS (basic service set), the LAN stations (STAs) communicate with each other without any supporting infrastructure, and the topology is useful for such applications as file sharing in a limited geographic area (like a conference room). The second topology makes use of an access point (AP), which allows the LAN stations in a BSS to connect to a backbone local area network (wireline or wireless LAN) or distribution system (DS), and to extend its reach beyond the range of the BSS. The stations within the BSS can still communicate with each other directly without involving the AP, but a station cannot use another station in the BSS to route packets to the AP. This topology with access points supports LAN operation over larger geographic areas (large buildings, campuses) through the use of multiple APs.

5.6.2.1 Physical Layer. For the direct sequence spread spectrum scheme in IEEE 802.11, the original data signal is modulated by means of a single predefined spreading code, (unlike multiple codes in the IS-95 CDMA cellular system). Knowing the spreading code, the receiver can recover the original signal by despreading the received signal. The processing gain (bandwidth of the spread spectrum signal) of 11 (10.4 dB) provides adequate protection against narrow-band interference and allows the available frequency band (83.5 MHz in the United States) to be segmented into a number of direct sequence center frequencies. IEEE 802.11 specifies 11 such frequencies.

In the frequency-hopping spread spectrum scheme, the frequency of data transmission is continuously changing (hopping) in time among the 79 frequencies specified in IEEE Standard 802.11. The fixed time interval over which data are transmitted on a given frequency is called the *dwell time*. Again, knowing the frequency-hopping pattern, the frequency synthesizer in the receiver can synchronize with the hopping pattern of the incoming signal and extract the original data. Three different sets of frequency-hopping sequences are defined, each set containing 26 hopping sequences. Use of different FH sequences ensures that adjacent or overlapping BSS can operate without interference and maximize throughput of each BSS. It is estimated that an FH scheme will provide better performance/capacity in a dense environment, containing many overlapping BSSs, and also in environments with high levels of narrowband interference.

5.6.2.2   MAC Layer.   The MAC layer is responsible for providing functions like channel allocation and access procedures, protocol data unit addressing, frame formatting, error checking, and packet segmentation and reassembly. The IEEE 802.11 MAC layer protocol is designed as a flexible protocol that can support the range of potential operating environment choices in terms of:

- physical layers (direct sequence or frequency-hopping spread spectrum)
- service types (asynchronous, time-bounded, contention-free)
- network topologies (access point, independent BSS)
- power management (with infrastructure, without infrastructure)

The primary access protocol used in the 802.11 MAC sublayer is called the distributed coordination function (DCF) and is based on carrier-sense multiple access with collision avoidance (CSMA/CA). It employs a *random backoff* mechanism to reduce the probability of collision between two frames. Thus, each station ready to transmit data invokes the random backoff when it senses the carrier status change from busy to idle. The random backoff time follows a uniform distribution measured in discrete time slot durations, and the maximum range of the distribution is called the *contention window* (CW). The CW parameter is increased geometrically after each consecutive collision. The occurrence of a collision is indicated by the lack of an acknowledgment (ACK) frame. Different levels of priority are assigned for transmission of ACK frames (highest priority), time-critical frames, and asynchronous data frames (lowest priority) by assigning appropriate *initial interframe spaces* (IFS) to these frames. The media access mechanism used in IEEE 802.11 is shown in Figure 5.18.

The point coordination function (PCF) is used in 802.11 to support time-bounded services whereby a point coordinator or PCF station can have priority control of the medium. This mechanism is implemented by the com-

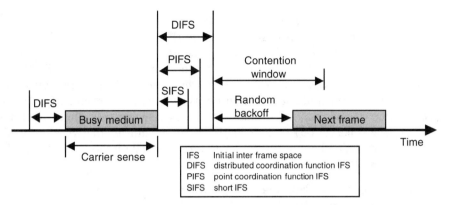

Figure 5.18 IEEE 802.11 MAC sublayer: medium access mechanism.

bined use of a point coordination function IFS (PIFS) and a beacon frame
that informs other stations in the cell to hold transmissions for the duration
of the contention-free period. Using an (optional) polling frame, the PCF
station can grant medium access to any specific station that has requested for
contention-free access. The contention-free period is adjustable to accommodate different load and congestion situations. Generally the access point
will act as a PCF station in an AP-supported topology.

The MAC sublayer also includes an option of using request to send (RTS)
and clear to send (CTS) frames to overcome the *hidden node* problem, which
occurs when two stations attempting to transmit to a receiver are unable to
hear each other's transmissions. The two transmitters therefore do not invoke
the random backoff, thereby increasing the probability of collision. The
RTS/CTS operation provides a handshake mechanism in which the potential
transmitting station issues an RTS frame to the receiving station, which responds with a CTS frame. The CTS frame, while granting permission to the
transmitting station, also acts as an indication to other stations within its range
to hold off transmissions for a given interval called *net allocation vector*
(NAV). To conserve signaling resources, the RTS/CTS feature is invoked only
when long packets are to be transmitted.

Since conservation of battery power in mobile stations is a critical factor,
IEEE 802.11 provides the MAC sublayer with special power management features for topologies with and without access points. In the former case, a mobile station in the sleep or power-conserving mode can wake up from time to
time and listen to selected beacon frames sent by the AP. If it detects a signal
indicating buffered data, it can request delivery by sending a polling frame. In
the case of infrastructure-free topology, a station in a sleep mode can wake up
for short predetermined intervals to monitor the carrier and decide whether
to continue in the wake-up state.

5.6.2.3   Authentication and Encryption.   The user/station authentication and encryption features provided in the IEEE 802.11 specification serve to protect the authenticity of the LAN user/station and the privacy of the data over the radio link. The authentication method proposed in IEEE 802.11 is based on *shared secret key* mechanisms, where the authentication keys are shared between the authentication authority and the user/terminal. These authentication mechanisms are similar to those used in current cellular systems. The privacy (encryption) feature in IEEE 802.11 utilizes the wired equivalency privacy (WEP) algorithm, which in turn is based on the RC4 PRNG algorithm from RSA Data Security.

5.6.3   High Performance European Radio LAN (HIPERLAN)

The starting objectives for the European wireless LAN specification were to provide performance (throughput, security) equivalent to a wireline LAN such as Ethernet, and to provide some support for isochronous services like video and image. Some additional requirements included seamless roaming, a

Table 5.1 System Parameters for HIPERLAN

System Parameter	HIPERLAN Value
Transmission frequency band	5.150–5.300 GHz[a]
Number of channels	5
Bandwidth/channel	23.5294 MHz
Mobility limit	< 5 km/h
Modulation	
High bit rate (HBR)	GMSK
Low bit rate (LBR)	FSK
Data rate (HBR transmission)	23.5294 Mb/s
Data rate (LBR header signaling)	1.47060 Mb/s
Maximum burst duration	1 ms
Maximum burst length each	47 blocks of 496 bits
Interleaving depth	16
Power level (maximum)	1 W
Error correction	BCH code $(31, 26, 3)$
Error detection	CRC (32 bits)

[a] CEPT has also identified the 17.1–17.2 GHz band for potential WLAN applications. However, HIPERLAN does not have exclusive use of the 5 and 17 GHz bands.

range exceeding 50 meters, and low power consumption. The HIPERLAN standard operates at 23.529 Mb/s and supports multihop routing, time-sensitive traffic, and power-conserving methods. The specification was developed by ETSI and is limited to the specification of the physical and data link layers, the latter being divided into the channel access control (CAC) and the medium access control (MAC) sublayers. The primary characteristics and system parameters for HIPERLAN are summarized in Table 5.1.

The MAC protocol features available in HIPERLAN include the following:

- service access point (SAP) interface
- efficient contention resolution protocol
- connectionless, peer-to-peer structure
- variable packet lengths
- support of time-bounded services and QOS for packet delivery
- use of *forwarders* and multihop routing
- support of *sleep* and *doze* modes for terminal power conservation

5.6.3.1   Channel Access and Contention Resolution.   The HIPERLAN
MAC protocol is based on a carrier-sensing or listen-before-talk mechanism. If a node senses the medium to be free for a time corresponding to 1700 bits, the node can transmit immediately. If not, the node enters a channel access cycle that consists of three phases: prioritization, elimination, and yield. The aim of the *prioritization phase* is to allow only nodes that have the highest priority data to contend for the channel. During the *elimination phase,* only the nodes that survived the prioritization phase are contending for the channel. Further elimination or screening is achieved by having each node transmit a burst whose duration is geometrically distributed (in terms of number of slots) and then listen for a time duration equal to one time slot. If another burst is detected, the node drops off from contention for the channel. In the *yield phase,* which immediately follows the *elimination phase,* a node that survived the *elimination phase* defers transmission for a geometrically distributed number of time slots while listening to the channel. If it senses the channel to be busy (indicating that another node has already gained access), the node drops off completely. The node can attempt channel access at the next cycle. The procedure is found to be quite effective, and the collision probability is estimated to be less than 3%.

5.6.3.2   Forwarders.   Forwarders are used in HIPERLAN to extend its
range through multihop routing. Specific nodes can announce themselves as forwarders and act as relay stations for routing packets in the chain. The forwarder nodes need to have the capability to maintain and update routing databases. The routing in HIPERLAN is based on two SAP addresses for the

MAC sublayer and for the channel access control (CAC) sublayer associated with each HIPERLAN MAC signaling data unit (SDU). For a single-hop connection, these addresses are identical. For multihop connections, the destination CSAP (channel access control sublayer SAP) determines the appropriate forwarder node, which is associated with the originating node. The MAC entity in the forwarder node looks up the new CSAP address from the routing tables and forwards the SDU onward in the multihop chain.

5.6.3.3 Power Conservation. HIPERLAN provides for terminal power conservation in two ways. In the *sleep* or *p-saver* mode the terminal is essentially disabled (very few active circuits) and awakes only at predetermined time instances. Associated with a *p-saver* node are *p-supporter* nodes, which know the sleep/awake pattern of the *p-saver* node(s) and initiate new or stored messages to the *p-saver* node(s) when the latter are in awake state. HIPERLAN also provides for a *doze* state in which only a simple, low power FM receiver in the node is enabled. The FM receiver (i.e., the *dozing* node), however, is capable of reading the low bit rate (LBR) header, which contains an abbreviated CSAP destination address. If the *dozing* node recognizes that the transmission is addressed to it, it activates the full high bit rate (HBR) decoder and powers up its receiving equipment.

5.7 SUPPORT OF MOBILITY ON THE INTERNET: MOBILE IP

With the rapidly increasing penetration of laptop computers, which are primarily used by mobile users to access Internet services like e-mail and the World Wide Web, support of mobility in the Internet is an emerging requirement. The key problem that mobile IP attempts to solve is the development of a mechanism that allows IP nodes to change their physical locations without having to change IP address, thereby providing so-called "nomadicity" to the Internet users. However, the Internet protocol and the OSI architecture are based on the assumption that the end systems are stationary. Thus, the optimum solution should on the one hand ensure that the mobile nodes can function efficiently in terms of receiving and transmitting data when they change their points of attachment, and on the other hand ensure minimum impacts on the Internet infrastructure and protocols, as well as on the host or end system software.

IP addresses are used for two distinct purposes in the Internet—for uniquely identifying the end system and for routing packets to the end system. In mobile IP, a mobile end system is assigned a second temporary IP address purely for routing packets to the end system when the mobile moves to another subnetwork. The mobile IP has been under development by the Internet Engineering Task Force (IETF), and the draft specification is available in the form of a request for comment (RFC). Mobile IP work in the IETF, in some sense, complements the work on wireless LANs in IEEE 802.11 and the HIPERLAN, whereby mobile IP extends the domain of mobility for wireless LAN users. This is illustrated in Figure 5.19.

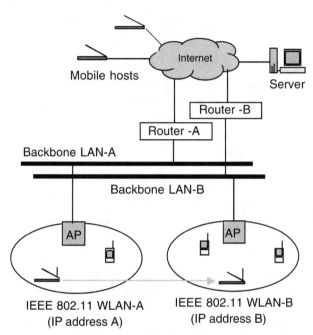

Figure 5.19 Illustration of WLAN and mobile IP application.

Mobile IP, which is supported in Internet Protocol Version 4 (IPv4), the current version, allows the end systems to change their point of network attachment without impacting the normal operation of the Internet. The basic architecture of mobile IP to accomplish this is illustrated in Figure 5.20.

The mobile IP architecture is based on defining and implementing two key functional entities; the home agent (HA) and the foreign agent (FA). The basic procedure in mobile IP requires the following steps:

Agent discovery. Mobility agents advertise their availability on each link through which they provide service in the coverage area. This allows a visiting mobile node (MN) to recognize that it is outside its home agent's area. The visiting MN communicates with the FA, and a *care-of* address is assigned. If the MN knows that it has moved out of its home network, it can use dynamic host configuration protocol (DHCP) or point-to-point protocol (PPP) to contact a service provider in the visited network and obtain a *care-of* address.

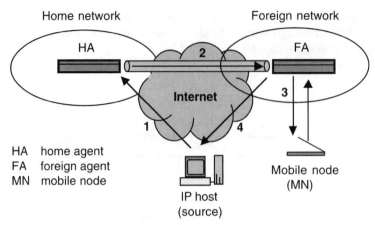

Figure 5.20 Mobile IP architecture and functional components.

Registration. Once a *care-of* address has been assigned to the visiting MN, it registers it with its HA by way of the FA. The HA then maintains a *binding* or association between the permanent IP address and the *care-of* IP address for routing and tunneling of datagrams addressed to the MN.

Tunneling. When the *care-of* address for the MN has been registered with the HA, it is ready to tunnel any datagrams addressed to the MN to its current FA, which is responsible for delivering them to the MN. Tunneling essentially involves encapsulation of the datagrams so that the permanent IP address of the MN is invisible to the transit routers and only the *care-of* address is displayed in the header field.

As illustrated in Figure 5.20, a datagram addressed to the MN is always routed to the HA, which tunnels it to the FA for delivery to the MN. The FA *detunnels* or *decapsulates* the datagram. Return packets, however, are not routed via the HA; rather, they are routed directly to the source node via standard IP routing procedures. Whereas agent discovery, registration, and tunneling are the three basic functions in mobile, IP, the support of these functions is distributed among a number of subtasks. Some of these subtasks are as follows:

Mobile agent discovery, which is similar to that used by Internet nodes to detect routers running ICMP (Internet control message protocol) router discovery.

Agent advertisement, which is an extension applied to the ICMP router advertisement procedure.

Agent solicitation, which uses ICMP router solicitation messages.

Registration request, which depends on whether the MN obtained the *care-of* address from an FA or from another independent service using dynamic home configuration protocol (DHCP). In the former case the MN sends the registration request to the FA, which relays it to the HA. In the latter case, the request is sent directly by the MN to the HA.

Registration reply, which confirms (or denies) the registration request and indicates the duration for which the registration will be honored. In case of registration request denial, cause for denial may be provided.

Securing registration procedure, which provides protection from fraudulent registration attempts. Three authentication extensions are defined to cover communications to and from MN and FH, MM and FA, and HA and FA.

Routing and tunneling, which is generally accomplished by using the default *IP-within-IP encapsulation* algorithm. Alternatively, the *minimum encapsulation* can be used, which requires fewer overhead bytes. Use of minimum encapsulation requires agreement between the MN, HA, and FA (if present).

A number of improvements to the current mobile IP (IPv4) are planned for the next version of the Internet Protocol (IPv6). One of the major drawbacks of the current mobile IP specification is that it requires all datagrams to the MN to be routed and tunneled via the HA, leading to the so-called *triangular routing*. This type of routing places an undue load on the HA, especially if a large number of nodes under its service area are roaming outside. The proposals for support of mobility within IPv6 include making all IPv6 nodes *mobile-aware* (so that datagrams to a roaming MN can be tunneled directly) and eliminating the need for FAs (by providing a method that allows the MNs to obtain *care-of* addresses by using the IPv6 neighbor discovery protocol).

5.8   MOBILE MULTIMEDIA: WIRELESS ATM (WATM)

Asynchronous transfer mode (ATM) is generally accepted as the platform for supporting end-to-end, broadband multimedia services with guaranteed quality of service (QOS). ATM networks are designed to efficiently handle various classes of service including both packet mode and synchronous services; they are scalable, and can support different transmission rates and environments. Wireless ATM (WATM) aims to provide an integrated architecture for seamless support of end-to-end multimedia services in the wireline as well as wireless access environments. Thus, WATM is expected to meet the needs of wireless users who are looking for a common networking solution that can meet their high speed data and multimedia services requirements with excellent reliability and service quality. Standardization activities on WATM have been underway in Europe (ETSI) and North America (ATM Forum) since 1996, and activities in this area have also started in Japan (ARIB). A number of WATM test beds have also been implemented in the United States and in

Europe to assess the feasibility and suitability of various implementation options. The results from these studies are being brought to the standards forums for discussion and decision.

Mobile networks in the current generation were primarily designed for low speed, circuit-switched services and may be classified as narrowband networks. The available data rates in these networks are inadequate for supporting high speed Internet connections and mobile multimedia services. The next generation of mobile systems, like IMT-2000 and UMTS, are expected to provide data rates up to 2 Mb/s and will represent a significant improvement for packet mode, high speed Internet access, but will still fall short of the ability to support full range of real-time multimedia applications. For applications of these types, the radio interface must provide channel data rates in excess of 2 Mb/s. To meet these high data rates, a band of 100 MHz or more would have to be assigned for the wireless broadband radio interface. This type of large bandwidth assignment cannot be achieved within the lower frequencies assigned for the second- and third-generation mobile systems (in the 800 or 2000 MHz band). The frequency bands being considered for mobile multimedia applications are therefore in the higher frequency bands. Frequencies in the 5, 17, 40, and 60 GHz bands are being considered for mobile multimedia services. Current focus of WATM activity is the assignment of 150 MHz in the 5 GHz band (5.15–5.30 GHz) for wireless LANs in the United States and Europe. The FCC in the U.S.A. has also identified an unlicensed national information infrastructure (U-NII) band, which provides 300 MHz in the 5 GHz band.

Many technical challenges associated with developing wireless ATM have their origin in the limitations of the current specifications for ATM, which were based on high speed, reliable (very low error rates) fiber-optic transmission links (some times even including the subscriber loop). The bandwidth (hence the available data rates) and the reliability of the wireless access links, however, are relatively much lower. To accommodate the special needs of the radio medium and the requirements for mobility, special design of the physical medium, medium access, and data link controls is required, as well as additional functionality in the ATM network for efficient mobility management. The design of the radio interface also needs to be optimized for smooth and hopefully transparent operation with the existing ATM data and signaling formats.

The WATM system architecture is centered around the standardized high speed ATM network signaling and control procedures providing multimedia services to wired and wireless terminals. A simplified architecture for wireless ATM is illustrated in Figure 5.21.

Beside the standard ATM network components used in wireline networks, the WATM architecture is based on the following additional components to meet the needs for mobility associated with wireless ATM terminals:

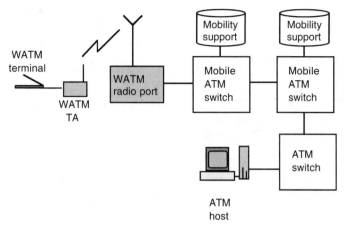

Figure 5.21 Proposed wireless ATM architecture.

- a WATM terminal adapter (TA) that supports new radio access layer (RAL) protocols

- a radio port (or base station) that interfaces the radio segment to the ATM network

- a mobile ATM switch that incorporates mobility management capabilities

As far as possible the WATM will utilize the standard interfaces defined for B-ISDN/ATM networks (e.g., Q.2931, B-ISUP, PNNI). However, modifications to these interfaces will be needed to support the mobility management and radio resource control requirements of wireless access. New data link control (DLC) and medium access control (MAC) layers need to be defined, not only to maximize the throughput across the radio link but also to ensure that information across the radio interface can easily be formatted into the standard cells used in the ATM network. The protocol architecture model for signaling (C plane) in WATM is shown in Figure 5.22.

In the model it is assumed that the WATM terminal adapter functions are integrated in the WATM terminal so that a separate interface between the terminal and the TA, some times referred as W-UNI, is not shown. The WATM TA provides necessary functions for mobility management (e.g., location management, handoff signaling, wireless link performance monitoring). Further, the WATM TA supports the so-called R functions, which include the termination of physical, MAC, and DLC layers, as well as wireless control (metasignaling). Essentially, WATM extends the capabilities of standard ATM functions in the following two areas:

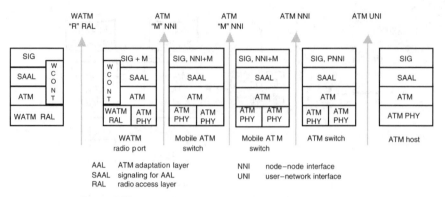

Figure 5.22 Protocol architecture model for WATM signaling.

- Radio access layer (RAL) protocols that support

  logical link control (LLC)

  medium access control (MAC)

  high speed radio physical layer (PHY)

  wireless control

- ATM network layer protocols for mobility that support

  Location management

  terminal/user registration (including authentication and privacy)

  handoff management

  routing and QOS management

5.8.1 LLC Sublayer

The primary function of the logical link control sublayer is to provide error control on a per-connection basis to satisfy the QOS requirements. This is achieved by deploying suitable ARQ procedures. Proper sequencing of the cells, which may be disrupted as a result of retransmissions, is also handled in the LLC. For maintaining the cell sequence, a header containing the packet sequence number is added to each ATM cell, and buffers are maintained at the receiving and transmitting ends for ensuring that cells within a given window are correctly received. Different services and applications in the ATM environment demand and are assigned different QOS parameters (delay, delay variation, cell loss probability, etc.). To efficiently transport multimedia traffic over the radio interface, while conserving limited radio resources, error control procedures must have sufficient flexibility to permit them to be tailored for each connection based on the exact QOS requirements.

5.8.2   MAC Sublayer

The media access control sublayer specifies access procedures that allow efficient sharing of the radio channel among multiple users while supporting the standard ATM services and traffic classes: that is, ABR (available bit rate), CBR (constant bit rate), VBR (variable bit rate), and UBR (universal bit rate), each with its associated QOS controls. The type of MAC protocol used depends on the radio transmission technology deployed at the physical layer. A variety of contention-free and contention-resolution-type protocols are being considered for WATM MAC sublayer specification. These proposals fall into two categories based on the physical layer design: spread spectrum (SS) CDMA-based methods and TDMA-based methods. The TDMA-based methods may be further subdivided into TDMA/FDD and TDMA/TDD.

CDMA techniques have been applied in cellular mobile networks very successfully both in the 800 and 1800 MHz bands with significant capacity gains. When this access method is applied to WATM, the entire medium (frequency band) is used by each of the connections (connections being differentiated only by their spreading codes or frequency-hopping patterns); there is no contention for the channel. In this case the MAC design is greatly simplified and is limited to providing such higher level functions as packet scheduling, admission control, and handoff management. However, the SS-based methods may have large bandwidth requirements for high bit rate data and multimedia applications. Some experimental WATM systems have deployed frequency-hopping spread spectrum methods.

In the case of TDMA channels, the MAC protocol is generally based on some variation on the slotted ALOHA technique for channel access request and time slot reservation in the uplink (mobile terminal-to-radio port) direction. In the downlink direction the radio port has complete control of the channel and utilizes time division multiplexing in a scheduled mode for sending acknowledgment (ACK) and information packets. In case of TDMA/FDD the assigned frequency is partitioned into uplink and downlink portions, and the two paired frequencies need sufficient separation to permit uplink and downlink transmissions to take place simultaneously, without interference. In the TDD duplexing, there is no pairing of the uplink/downlink frequencies and only one carrier frequency is used in both directions. Since the transmissions in the two directions are separated in time, some extra delay is added between the transmission and reception modes. For example, a mobile terminal may encounter some delay in receiving an ACK from the radio port, as opposed to almost instantaneous response in FDD. A number of specific MAC protocols for TDMA/TDD and TDMA/TDD operation are under consideration. The descriptions of these methods along with a comparison of their performance characteristics were published late in 1997 [33].

5.8.3   Physical Layer

As mentioned in connection with the MAC layer, the possible multiple access techniques for WATM include direct sequence (DS) CDMA, frequency hopping (FH) CDMA, TDMA/FDD, and TDMA/TDD. The WATM physical layer needs to deploy modulation and encoding methods that can provide high spectrum efficiency and deliver data rates in excess of 10 Mb/s. Both single-carrier and multicarrier modulation schemes are being studied. The multicarrier modulation that seems to be a front-runner is orthogonal frequency division multiplex (OFDM). It has the advantage of not requiring equalizers, and it uses extended transmission symbol periods and interleaving of ATM cells across the carrier multiplex to provide good protection against multipath effects. However, OFDM requires complex frequency transformations and, because of multicarrier operation, can introduce delays. Some of the other options being considered include QPSK/QAM, offset quaternary phase shift keying (OQPSK), and multilevel QAM (16QAM) with code division multiple access. Some form of forward error correction like Reed–Solomon encoding is also needed.

The packet format over the radio interface needs to be optimized for maximum throughput. Most approaches recommend header compression to reduce ATM header size on the radio link. Generally the 5 byte ATM header is compressed to 2 bytes, which carries the virtual circuit identifier (VCI) and minimum ATM control information.

5.8.4   Wireless Control

The wireless control sublayer resides in the WATM TA and the radio port and provides control plane functions for the integration and interworking of the radio access layer and the ATM network. Wireless control is essentially a management sublayer for the physical, MAC, and LLC sublayers. It manages terminal migration/location, terminal registration/authentication, and handoff control. Additional functions this management sublayer may undertake include power measurement and control, various stages of handoff management, and connection state transfers between radio ports.

5.8.5   Mobility Management

Specification of mobility management procedures at the network level will require significant extensions to existing ATM architectures and protocols—specifically, for the radio port controller and the mobile ATM switch. The basic procedures for location management, terminal/user registration, and authentication and encryption are well defined and extensively implemented in existing cellular mobile networks. To a large extent, therefore, the specification of these functions can be based on existing procedures in (say) GSM or

ANSI-41 networks. The support for handoffs will have a significant impact, however, because it will require extensions to existing ATM routing algorithms. Every request for a handoff will invoke route changes and optimization in real time, In other words, every handoff event during a multimedia call will result in a significant change in the optimal route of each virtual circuit (VC) involved in the multimedia call. Most of the VC rearrangements or reassignments will be in the vicinity of the WATM terminal (which is requesting handoff), and the reassignments need to be optimized for cost and performance (QOS). Because the rearrangements for handoffs is a time-critical task, it may be expedient to use a simple path extension scheme followed (later) by a more optimized arrangements of the connections.

5.9   CONCLUDING REMARKS

With the rapidly increasing usage of portable laptop computers and growing user demand for Internet access services, the need for high speed mobile data communications is coming into greater focus. This chapter has summarized the data communication capabilities currently available on the cellular mobile networks and on other emerging technologies such as high speed LANS, mobile IP, and wireless ATM. Besides the mobile data communication capabilities of these systems, a large mobile data communications market is served by the private packet radio, private mobile radio, and specialized mobile radio industry as well as by wireless PBX-type systems (CT2, DECT, PACS, PHS) and satellite systems. Whereas most existing satellite systems provide connection-oriented channels for data transport, proposals like Teledesic intend to use a constellation of low earth orbit (LEO) satellites to start providing high speed backbone connectivity for multimedia applications within the next few years.

5.10   REFERENCES

[01] N. J. Muller, *Wireless Data Networking,* Artech House, Boston, 1995.

[02] K. Apostolis et al., "Mobile Packet Data Technology: An Insight into MOBITEX Architecture," *Personal Communications Magazine,* Vol. 4, No. 1, February 1997.

[03] I. Channing, "TETRA-Digital PMR for the 21st Century". *Mobile Communications International,* Issue 53, July/August 1998.

[04] P. Harley, "The TETRA Standard," *Australian Telecommunications Journal,* Vol. 48, No. 4, 1998.

[05] J. Jayapalan et al., "Cellular Data Services Architecture and Signaling," *Personal Communications Magazine,* Vol. 1, No. 2, 1994.

[06] J. Hamalainen, "High Speed Data Service in GSM," Technical Forum, Telecom '95, Geneva, October 1995.

[07] C. Scholefield, "Evolving GSM Data Services," IEEE International Conference on Universal Personal Communications, San Diego, CA, September 1997.

[08] CDPD Forum, "Cellular Digital Packet Data Specification—Release 1.1," Technical Report, CDPD Forum, Chicago, January 1995.

[09] K. C. Budka et al., "Cellular Digital Packet Data Networks," *Bell Labs Technical Journal,* Vol. 2, No. 2, 1997.

[10] Y.-M. Chuang et al., "Trading CDPD Availability and Voice Blocking Probability in Cellular Networks," *IEEE Network,* Vol. 12, No. 2, March/April 1998.

[11] D. Saha et al., "Cellular Digital Packet Data Network," *IEEE Transactions on Vehicular Technology,* Vol. 46, No. 3, August 1997.

[12] European Telecommunications Standards Institute, "Digital Cellular Telecommunication System (Phase 2+); General Packet Radio Service (GPRS); Overall Description of the GPRS Radio Interface; Stage 2 (GSM 03.64 version 5.1.0)," TS 03 64 V5.1.0, ETSI, Sophia Antipolis, France, 1997.

[13] S. Hoff et al., "A Performance Evaluation of Internet Access via the General Packet Radio Service in GSM," IEEE Vehicular Technology Conference, Ottawa, Canada, May 1998.

[14] C. Scholefield, "Evolving GSM Data Services," IEEE International Conference on Universal Personal Communications, San Diego, CA, 1997.

[15] J. Cai and D. Goodman, "General Radio Packet Service in GSM," *Communications Magazine,* Vol. 35, No. 5, October 1997.

[16] M. Onuki et al., "Mobile Packet Data Communication in a TDMA Cellular System," IEEE International Conference on Universal Personal Communications, Cambridge, MA, September 1996.

[17] S. Hirata et al., "PDC Mobile Packet Data Communication Network," IEEE International Conference on Universal Personal Communications, Tokyo, October 1995.

[18] European Telecommunications Standards Institute, "EDGE Feasibility Study, Work Item 184; Increased Data Rates Through Optimized Modulation," Technical Document SMG2 95/97 Version 0.3, ETSI, Sophia Antipolis, France, December 1997.

[19] A. Furuskar et al., "System Performance of EDGE, a Proposal for Enhanced Data Rates in Existing Digital Cellular Systems," IEEE Vehicle Technology Conference, Ottawa, Canada, May 1998.

[20] C. J. Fenton, et al., "Mobile Data Services," *BT Technical Journal,* Vol. 14, No. 3, July 1996.

[21] P. Schramm et al., "Radio Interface Performance of EDGE, a Proposal for Enhanced Data Rates in Existing Digital Cellular Systems," IEEE Vehicular Technology Conference, Ottawa, Canada, May 1998.

[22] D. J. Harasty, L.F. Chang, and A. R. Noerpal, "Architecture Alternatives for Wireless Data Services," International Conference on Universal Personal Communications (ICUPC '94), San Diego, CA, September/October 1994.

[23] H. Yoshida et al., "PHS Packet Data Communications System," *NTT Review,* Vol. 9, No. 3, May 1997.

[24] Institute of Electrical and Electronics Engineers, "Wireless LAN Medium Access Control (MAC) and Physical Layer (PHY) Specifications," P802.11D3, IEEE Standards Department, Piscataway NJ, January 1996.

[25] B. P. Crow et al., "IEEE 802.11 Wireless Local Area Networks," *Communications Magazine,* Vol. 35, No. 10, September 1997.

[26] R. O. LaMaire et al., "Wireless LANs and Mobile Networking: Standards and Future Directions," *Communications Magazine,* Vol. 34, No. 8, August 1996.

[27] A. Wittneben and W. Liu, "The European Wireless LAN Standard HIPERLAN: Key Concepts and Testbed Results," IEEE Conference on Vehicular Technology Conference, Phoenix, AZ, May 1997.

[28] C. Perkins, "IPv4 Mobility Support," Internet Engineering Task Force, RFC 2002, IETF, Reston, VA, October 1996.

[29] C. Perkins, "Mobile IP," *Communications Magazine,* Vol. 35, No. 5, May 1997.

[30] D. Raychaudhary, "Wireless ATM Networks: Architecture, System Design and Prototyping," *Communications Magazine,* Vol. 34, No. 8, August 1996.

[31] A. Acharya et al., "Mobility Support for IP over Wireless ATM," *Communications Magazine,* Vol. 36, No. 4, April 1998.

[32] J. Mikkonen et al., "Emerging Wireless Broadband Networks," *Communications Magazine,* Vol. 36, No. 2, February 1998.

[33] J. Sanchez et al., "A Survey of MAC Protocols Proposed for Wireless ATM," *IEEE Network,* Vol. 11, No. 6, December 1997.

CHAPTER 6

IMT-2000: THIRD-GENERATION MOBILE COMMUNICATION SYSTEMS

6.1 INTRODUCTION

6.1.1 Background

Current forecasts indicate that demand for wireless access to global telecommunications will reach 1 billion users by year 2010, exceeding the likely number of wired or fixed access lines. Further, if Internet access, which has been doubling every year since 1988, continues to grow at this level, the existing fixed network will be eclipsed in the very near future. The emerging Internet environment urgently requires support for asymmetric, interactive, multimedia traffic based on high speed packet data transport. Such rapidly growing service requirements, driven by the global users of telecommunications, will dramatically change the nature of telecommunication services and the underlying networks in the twenty-first century.

The underlying vision for the emerging mobile and personal communication services for the new century is to enable communication with a person, at any time, at any place, and in any form, with a paradigm shift from the current focus on voice and low speed data services to high speed data and multimedia services. The current second-generation digital mobile and personal communication systems are based on national or regional standards that are optimized for region- or country-specific regulatory and operating environments. They are therefore unable to interoperate with each other and can provide mobility only within their radio environments as well as within geographic regions in which a specific standard is operational.

Efforts are therefore under way at the international as well as the regional/national levels to define the so-called *third-generation* mobile telecommunication system that will meet the coming needs of telecommunications subscribers. It is well recognized that international or global standards for mobile telecommunications are needed, not only to ensure seamless global mobility and service delivery but also for integrating the wireline and wireless network to provide telecommunications services transparently to the users. These global standards must be flexible enough to meet local needs and to allow current regional/national systems to evolve smoothly toward the third-generation system.

The International Telecommunication Union (ITU), the United Nations organization responsible for global telecommunications standards, has been working since 1986 toward developing an international standard for wireless access to worldwide telecommunication infrastructure. This standard is known as IMT-2000, for International Mobile Telecommunications 2000, where *2000* indicates the target availability date (year 2000) as well as the operational radio frequency band (2000 MHz range) for the standard. Until 1997, IMT-2000 was known as Future Public Land Mobile Telecommunication Systems (FPLMTS).

IMT-2000 is intended to form the basis for third-generation (3G) wireless systems, which will consolidate today's diverse and incompatible mobile environments into a seamless radio and network infrastructure capable of offering a wide range of telecommunication services on a global scale. Within the ITU, the *radio aspects* for IMT-2000, especially the selection of radio transmission technology (radio interface) and spectrum usage, are addressed in the Radio Communication Sector (ITU-R), whereas the *network aspects*, which include definition of network signaling interfaces, services, numbering and identities, quality of service, security, and operations and management for IMT-2000, are addressed by the Telecommunications Standardization Sector (ITU-T). The specifications are captured in the so-called ITU Recommendations (voluntary standards), which provide the essential backbone for worldwide telecommunications. Work is also under way in regional/national standards forums like ETSI (Europe), the TIA and T1 committees (North America), and TTC (Japan) on third-generation wireless systems that complement and provide inputs and direction to the IMT-2000 activities in the ITU.

6.1.2   IMT-2000 Vision

The vision for an IMT-2000 system and its capabilities is summarized in Figure 6.1, which illustrates that IMT-2000 will provide capabilities constituting significant improvements over the current mobile systems, especially in terms of global mobility for the users and support of services like high speed data, multimedia, and Internet. Since, however, it is generally accepted that these IMT-2000 capabilities will to a large extent be achieved by evolving existing wireless and wireless networks, IMT-2000 will be a *family of systems* rather than a single, monolithic network.

In scope, IMT-2000 service environments will address the full range of mobile and personal communication applications shown in Figure 6.2: in-building

IMT-2000 Vision

- Common spectrum worldwide (1.8–2.2 GHz band)

- Multiple radio environment s (cellular, cordless, satellite, LANs)

- Wide range of telecommunications services (voice, data, multimedia, internet)

- Flexible radio bearers for increased spectrum efficiency

- Data rates up to 2 Mb/s (phase 1)—for indoor environments

- Maximum use of IN capabilities (for service provision and transport)

- Global seamless roaming

- Enhanced security and performance

- Integration of satellite and terrestrial systems

> **A global standard to satisfy market demand for mobile services in the 21st century**

Figure 6.1 Vision for IMT-2000 services and capabilities.

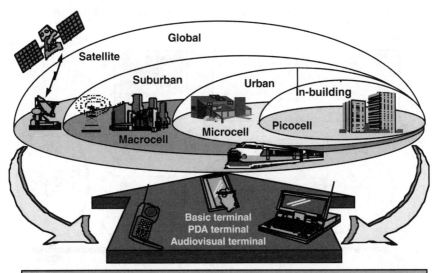

Figure 6.2 Scope of mobile services environments for IMT-2000.

(picocell), urban (microcell), suburban (macrocell), and global (satellite), as well as communications types that include voice, data, and image. Support of communication needs for developing countries in the form of fixed wireless access (FWA) applications is also included in the scope of IMT-2000.

6.1.3 IMT-2000 Evolution Aspects

Whereas the vision for IMT-2000 implies significant departures from the second-generation mobile and personal communication systems in terms of range of environments and services, as well as seamless global mobility, it is expected that IMT-2000 systems will essentially evolve from existing wireless and wireline systems as illustrated in Figure 6.3.

To provide required spectrum efficiency and capacity, the choice of the radio interface or radio transmission technology (RTT) for IMT-2000 may require a step change from the radio transmission technologies deployed in the current wireless/PCS systems. A number of radio transmission technologies have been developed for IMT-2000 in various parts of the world and were submitted to ITU-R for evaluation and for final selection by the end of 1999. The objective of ITU-R TG8/1 is to select RTTs that will cover all the radio operating environments with maximum commonalty. A brief description of some key radio technologies being evaluated by ITU-R is provided in Section 6.2.

The IMT-2000 networks, on the other hand, will need to evolve from the existing wireless and wireline networks, because it is not economical to replace existing network infrastructures by a completely new one. The pace of such an evolution will be determined by factors related to market demand, technology,

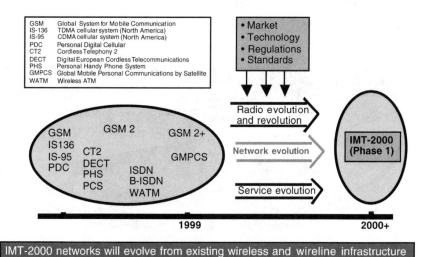

Figure 6.3 Expected evolution scenario for IMT-2000.

regulation, and standardization. The IMT-2000 *family of systems* concept adopted in the ITU-T reflects the industry view that there will be multiple IMT-2000 systems, each one being a member of the IMT-2000 *family* (e.g., GSM-based family members, ANSI-41-based family members). Whereas systems belonging to *different* family member groups will interoperate via ITU-defined interfaces to provide seamless global roaming and service delivery, systems belonging to the *same* family member group will continue to interoperate via interface definitions developed by the relevant regional/national standards forum. Progress toward definition of such interfaces in the ITU-T is summarized in Section 6.3.

6.2   IMT-2000 RADIO ASPECTS

6.2.1   Radio Spectrum for IMT-2000

Activities on IMT-2000 were initiated by the ITU based on a strong desire by the ITU membership administrations to develop worldwide standards for a global mobile telecommunication system. To achieve this objective, availability of a common worldwide frequency spectrum was essential. Based on detailed studies within ITU-R, the World Administrative Radio Conference (WARC), the body responsible for allocation and administration of radio frequencies at the international level, identified 230 MHz of global frequency spectrum for IMT-2000 at its 1992 meeting (WARC-92). This allocation also included spectrum for the satellite component of IMT-2000. The frequency in the 2000 MHz band identified for IMT-2000 by WARC-92 is as follows:

Uplink   1885–2025 MHz (1980–2010 MHz for mobile satellite service)
Downlink   2110–2200 MHz (2170–2200 MHz for mobile satellite service)

Some modifications in the allocations for the mobile satellite service, specifically for ITU Region 2 (Americas and Caribbean), were subsequently made by WARC-95. According to these modifications, Region 2 allocation for mobile satellite service (MSS) will be 1990–2025 MHz (uplink) and 2160–2200 MHz (downlink).

The satellite component of IMT-2000 spectrum would be subject to the international regulatory procedures for frequency coordination and registration with the ITU. These procedures are outlined in the ITU radio regulations and relevant World Radio Conference (WRC) resolutions, which have treaty status, representing international law to be observed by the signatory states. However, in accordance with prevailing practice, the actual allocation, utilization, and administration of the designated frequency spectrum for terrestrial services is a national matter, and in some countries/regions the lower part of the IMT-2000 band (1885–2025 MHz) is already being used for second-generation mobile/PCS systems. For example, DECT (Europe) operates at 1880–1900 MHz and PHS (Japan) at 1895–1918.1 MHz; the PCS systems based on the North American standards utilize 80 MHz duplex operation in the 1850–1990 MHz band. Further, existing terrestrial fixed services (e.g., microwave communication systems for utilities) in some countries are currently using the bands identified for IMT-2000 implementation. The current status of potential assignment of the IMT-2000 band of frequencies in key regions is shown in Figure 6.4.

The allocation of IMT-2000 frequency spectrum by WARC-92 was based on forecasts for third-generation wireless services that have turned out to be rather pessimistic. Recent forecasts for worldwide demand for wireless access in the early part of the twenty-first century are much higher and range from close to 1 billion mobile/PCS subscribers in year 2005 to more than 2 billion in year 2015, with majority of growth in the Asia–Pacific region. Further, the goals and objectives for IMT-2000 have been refined since 1992, and there now appears to be a growing demand for IMT-2000 to provide high bit rate data and multimedia services. Based on the revised forecasts, it is generally accepted that additional spectrum will be needed, and proposals for additional spectrum are being submitted to WRC-2000. For example, the European studies by the UMTS Forum have concluded that Europe will need the full 155 MHz available for third-generation terrestrial services by year 2005 and will reqire an additional 185 MHz by year 2010 to meet the market demand for terrestrial services. Similarly, for the satellite component it is estimated that Europe will need the full 60 MHz identified

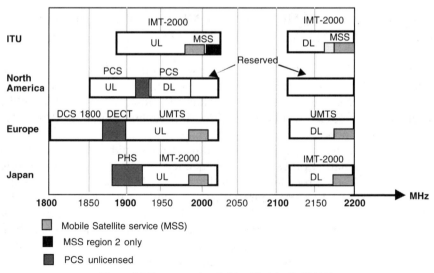

Figure 6.4 Frequency bands identified for IMT-2000.

by WARC-92 by year 2005, with an additional requirement for 30 MHz by year 2010. A consolidated view on worldwide spectrum requirements for IMT-2000 is being addressed by the ITU-R TG8/1 for submission to WRC-2000.

IMT-2000 will be required to support a wide range of traffic types including high speed packet data and multimedia. Traffic characteristics of such services differ significantly from telephony in that the demand over time varies in its request for speeds of transfer, length of transmission, streams, and symmetry of bidirectional transmissions. Sections 6.2.1.1 to 6.2.1.3 discuss some issues that further impact the spectrum requirements for IMT-2000 and are under study in the ITU as well as in the regional/national forums.

6.2.1.1   Transmission Modes.   Terrestrial wide-area systems like cellular mobile and the satellite applications will require paired bands for frequency division duplex (FDD) transmission, but short-range systems used for indoor and pedestrian-type applications can use unpaired asymmetrical band with time division duplex (TDD) transmission. To prevent traffic asymmetry and the resultant loss in spectrum efficiency, an optimal combination of FDD and TDD transmissions must be utilized. To maximize system capacity while maintaining flexibility, the exact split between paired and unpaired band spectra needs to be considered carefully.

6.2.1.2   Duplex Direction.   As indicated in Figure 6.4, the lower band of frequencies is identified for uplink (mobile to base station) and the upper band for the downlink (base station to mobile). This is considered to be the normal FDD duplex direction for the following reasons:

- It relaxes the transmit power requirements (hence increases battery life) for mobile stations because assigning the lower frequency provides a lower propagation loss.

- The commonality of the frequency usage plans between the satellite component and the FDD-based terrestrial component of IMT-2000 greatly facilitates the use of dual-mode mobile terminals supporting both satellite and terrestrial communications.

The reverse assignment proposal (lower band for downlink and upper band for uplink) is meant to facilitate migration from the PCS systems in North America, which are already using a significant part of the (lower) IMT-2000 band. However, there is currently no consensus around the proposal that reversing the duplex direction will lead to a more flexible and efficient operation.

6.2.1.3   Spectrum Efficiency and Sharing.   A significant element of IMT-2000 is the need to achieve a major improvement in spectrum efficiency compared to that currently available in the second-generation mobile communication systems. Four main factors affect spectrum efficiency:

radio transceiver technology, which includes access technology, modulation and coding, adaptive interference management, diversity techniques, and smart antenna technology

applications and services technology, including the use of packet transmission, asymmetry management, and data compression techniques

traffic management via delay management and the use of tariffs to reduce peak-to-average traffic ratios

radio channel access management by real-time management of access to the spectrum (to maximize spectrum utilization during peak traffic loads)

In the context of managing the access to the spectrum, sharing a common pool of spectrum between operators and/or between terrestrial and satellite services has been suggested as a method for improving spectrum efficiency. Sharing the same frequency spectrum between operators will lead to higher trunking efficiency and reduction in guard bands. However, these advantages need to be weighted against the air time overheads (for call setup, cleardown, handoff, etc.), as well as against increased complexity and cost of system design. Spectrum sharing between satellite and terrestrial services is generally not considered to be feasible because of the wide differences in received power flux density and transmitted power levels between the terminals operating in these two environments.

6.2.2   Radio Transmission Technologies (RTTs) for IMT-2000

6.2.2.1   General Requirements for Radio Access to IMT-2000. As opposed to second-generation wireless systems, which are generally optimized for circuit-switched services in a single radio operating environment, IMT-2000 is intended to provide access, by means of one or more radio links, to a wide range of telecommunication services (from voice to multimedia and Internet), operating in a variety of radio environments (e.g., see Figure 6.2). The presence of a wide variety of radio operating environments translates into a number of factors that impact the choice of candidate(s) for radio access technologies. Some examples of these factors are as follows:

- speed of movement (from zero to very high)
- user density (city centers to remote areas)
- physical environment (indoor, urban, suburban, rural, maritime, aeronautical)
- coverage (continuous or islands)
- delivery mode (terrestrial or satellite)

Another basic requirement is the need for IMT-2000 terminals to be able to roam globally and receive telecommunications services offered by IMT-2000 networks. Key features of the radio access for IMT-2000 therefore include the following:

- high level of flexibility
- cost-effectiveness in all operating environments
- commonalty of design worldwide
- operation within the designated MT-2000 frequency bands

6.2.2.2   *Evaluation and Specification Process for IMT-2000 RTTs.* ITU-R TG8/1 is responsible for specification of one or more radio transmission technologies for IMT-2000. Given the variety of radio operating environments to be supported by IMT-2000, ITU-R TG8/1 envisages that a set of RTTs (SRTTs) with maximum commonality across the components of the set will be required. ITU-R TG8/1 is aiming towards a flexible radio standard with multiple radio technology options. The timetable for initial (Release 99) specification of RTTs for IMT-2000 is shown in Figure 6.5.

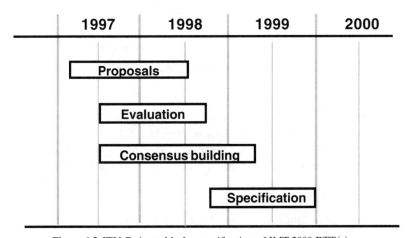

Figure 6.5 ITU-R timetable for specification of IMT-2000 RTT(s).

Based on a formal request to all its member administrations for submission of candidate radio transmission technologies for IMT-2000, ITU-R TG8/1 received 15 RTT proposals—10 of the proposals address terrestrial radio environments; the others address satellite applications. Proposals for the terrestrial environments received by the cutoff date (June 30, 1998) are summarized in Table 6.1.

A number of ITU-R member organizations representing various regions and countries are using the technical criteria chosen by ITU-R TG8/1 to evaluate the proposal submitted. Of course, many market- and business-

Table 6.1 RTT Proposals for IMT-2000 Terrestrial Component (submitted to ITU-R in June 1998)

Proposal	Source	Description
W-CDMA (UTRA)	ETSI (Europe)	Wideband CDMA selected by ETSI for UMTS terrestrial radio access
W-CDMA	ARIB (Japan)	Wideband CDMA selected by ARIB very similar to ETSI proposal
cdma2000	TIA TR45.5 (U.S.A.)	Wideband CDMA evolved from the IS-95 CDMA standard
WIMS W-CDMA	TIA TR46.1 (U.S.A.)	Wireless multimedia and messaging services wideband CDMA
NA: W-CDMA	T1P1-ATIS (U.S.A.)	Wideband CDMA closely aligned with the UTRA proposal
CDMA I	TTA (South Korea)	Multiband synchronous DS-CDMA
CDMA II	TTA (South Korea)	Asynchronous DS-CDMA
TD-SCDMA	CATT (China)	Time division synchronous CDMA
UWC-136	TIA TR45.3 (U.S.A.)	Universal wireless communications evolved from IS-136 (AMPS-TDMA)
DECT[a]	ETSI DECT project	Digital enhanced cordless telecommunications

[a] This proposal applies only to indoor and pedestrian radio operating environments.

related factors, in addition to purely technical considerations, will impact the final choice. The choice of the technical criteria is intended to provide a good assessment of the IMT-2000 system aspects that are dependent on radio transmission technologies. These technical criteria include the following:

- spectrum efficiency
- technology complexity—cost of installation and operation
- quality of service
- flexibility
- impact on network interfaces
- performance and capacity impacts on handheld terminals
- coverage efficiency
- power efficiency

The objective of the complex evaluation and *consensus building* process is to encourage discussion among the proponents of different RTTs to focus on the best features of their respective proposals and seek consensus toward a single, harmonized set of RTTs to be included in the IMT-2000 radio specifications. This process is essential because it is unlikely that any one of the proposed RTTs will be optimal for all possible combinations of IMT-2000 radio operating environments and meet the needs of different regions.

6.2.2.3   Characteristics of some Key RTT Proposals for IMT-2000.
As Table 6.1 indicates, some variation of the CDMA access method is being proposed for the IMT-2000 terrestrial component from various parts of the world. Although there is significant harmonization and alignment between some of these W-CDMA proposals (e.g., proposals from ETSI, ARIB, and T1P1-ATIS), at this stage there are significant differences between the various W-CDMA proposals to allow for a single RTT specification for the IMT-2000 terrestrial component. Table 6.2 compares major characteristics of the wideband CDMA proposals from Europe (ETSI), Japan (ARIB), and the United States (TIA TR45.5).

One of the key underlying principles in the development of the cdma2000 proposal from TIA TR45.5 was the need for high level of compatibility with and smooth migration from the IS-95 CDMA standard. Development of the ETSI/ARIB proposals had no equivalent constraints, which explains some of the differences in the system parameters. The key areas of difference include the chip rates, the need (or otherwise) for inter-BS synchronization, and the structure of pilot channels.

There has been a concerted effort by the industry to harmonize the various CDMA RTT proposals which has lead to an emerging consensus that the IMT-2000 radio technology specification from ITU-R TG8/1 should provide for a single, flexible standard that will support multiple modes including the following harmonized CDMA modes:

1. CDMA/FDD Direct Sequence (based on harmonized WCDMA proposals);

2. CDMA/FDD Multi Carrier (based on cdma2000 proposal); and

The harmozined chip rate proposed for CDMA/FDD Direct Sequence mode is $0.96 \times 4 = 3.84$ Mcps (as opposed to 4.096 Mcps in the original proposals), whereas the chip rate of $1.2288 \times 3 = 3.6864$ Mcps is retained for the CDMA/FDD Multi Carrier mode. Consensus has also been achieved for common pilots and common channels for the CDMA/FDD Direct Sequence option. Efforts are underway to harmonize the RF parameters between the CDMA modes in ordrer to achieve economy of scale in RF component design.

In addition to the above CDMA based radio technologies, the IMT-2000 radio interface standard will also include a TDMA/FDD mode based on the

Table 6.2 Comparison of Parameters[a] for three W-CDMA RTT Proposals

Parameter	W-CDMA (Europe)	W-CDMA (Japan)	cdma2000 (U.S.A.)
Multiple access	WB DS-CDMA[b]	WB DS-CDMA	WB DS-CDMA
Duplex method	FDD/TDD	FDD/TDD	FDD
Channel Bandwidth[c]	1.25/5/10/20 MHz	1.25/5/0/20 MHz	1.25/5/10/20 MHz
Chip rate (Mcps)[c]	$1.024 \times (1, 4, 8, 16)$	$1.024 \times (1, 4, 8, 16)$	$1.2288 \times (1, 3, 6, 12)$
Frame length	10 ms	10 ms (optional 20 ms)	20.5 ms
Inter-BS synch	Asynch	Asynch/synch	Synch
Data modulation FL/RL	QPSK/BPSK(FDD) QPSK/QPSK(TDD)	QPSK/BPSK(FDD) QPSK/QPSK(TDD)	QPSK/BPSK
Spread modulation FL/RL	QPSK/QPSK	QPSK/QPSK	QPSK/QPSK
Multirate concept	VSF + Multicode + multislot (TDD)	VSF/Multicode + multislot (TDD)	VSF/Multicode
Tx power control FL/RL	CLPC, 1.6 ks/s OLPC, 1.6 ks/s	CLPC, 1.6 ks/s OLPC, 1.6 ks/s	CLPC, 0.8 ks/s OLPC, 0.8 ks/s
Spreading codes	Short/long	Short/long	Short/long
Coherent detection FL/RL	With pilot symbol/ with pilot symbol	With pilot symbol with pilot symbol	With pilot channel with pilot symbol
Voice codec	Variable or fixed rate	Variable or fixed rate	Variable rate (EVRC)

[a] Parameter values in boldface indicate parameters that will be used in initial implementations.

[b] The ETSI and ARIB proposals have a TDD component for unpaired bands based on a hybrid W-CDMA + TDMA multiple access.

[c] In phase 1 (year 2001) user data rates up to 384 kb/s are expected to be supported so that the channel spacing of 5 MHz (and associated chip rates) will apply.

UWC-136 proposal and a TDMA/TDD mode based on the DECT proposal. Additionally, the radio specification may also include a hybrid TD-CDMA mode.

If the IMT-2000 users are to roam seamlessly among IMT-2000 systems that may deploy different radio technology and core network options, development of intelligent multimode terminals will be required. These multimode terminals should be able to determine the radio and network environment in which it is operating and select the appropriate mode transparently and automatically. A common protocol mechanism which will allow the terminals to automatically switch back and forth between different radio technology and different network protocol options is also being developed in the ITU.

6.2.3   Global Radio Control Channel

Considering that in the foreseeable future IMT-2000 will coexist with second-generation systems, there will be a large number of different standards and frequency bands for mobile communication networks. It is also expected that the RTT specification for IMT-2000 will include more than one RTT. It is expected that the core spectrum for IMT-2000 will be available and used worldwide, with some regional variations (Figure 6.4). However, extension bands may be different in different regions, especially considering that in the long term, the frequency bands used in the second-generation systems will gradually become available for the third generation.

In practice, a multimode terminal used by an IMT-2000 international roamer in a visited IMT-2000 network will have to scan for a suitable frequency band/channel, identify the applicable radio and network standard, and select from among the set or available services. If it develops that a very large number of frequency bands need to be scanned and the many standards need to be searched, registering such a roaming multimode terminal by means of a systematic scanning procedure will become very inefficient, tending to degrade the quality of service from the users' perspective. This problem could be alleviated by using a common physical or logical broadcast channel, called the *global radio control channel* (RCC), to scan a single frequency or a small range of frequencies, and thereby find the required information on available networks/standards and services.

WRC-2000 has on its agenda an item on identification of a global radio control channel to facilitate multimode terminal operation and worldwide roaming of IMT-2000. The basic requirement for the global RCC is cost-effective information transfer to the terminal, to facilitate roaming between networks supporting differing RTTs or sets of RTTs with differing spectrum utilization implementations. The mobile terminal may scan the global RCC at the time the terminal is switched on, on a periodic basis, or manually when deemed necessary. Two possible implementations for the global radio control channel are possible—a physical radio control channel and a logical radio control channel.

In the physical RCC implementation, an internationally standardized radio control channel will be broadcast in all regions where IMT-2000 systems are implemented. A roaming mobile can reference this global broadcast channel to obtain available types of alternative radio access scheme (RTTs, SRTTs), as well as pointers to their spectral locations, and then make a selection based on its own capabilities. The global channel may also provide information on the available networks and service capabilities. The advantages of this implementation are the ease of implementation and the flexibility available to the operators for the actual implementation. These benefits need to be weighed against the loss of usable spectrum (designated frequency or frequencies by the WRC), the need for coordination and ITU declaration, and the possibility of interoperator interference.

The logical global radio control channel is an alternative concept to the physical RCC implementation, where the RCC is implemented entirely as a *voluntary* logical channel within all public IMT-2000 networks. The logical RCC is used to broadcast information similar to the physical RCC implementation (RTTs and SRTTS with their spectral pointers), including information on networks and their service capabilities. This information is broadcast periodically (during non-peak-load periods) via a bulletin-type logical channel so that a user's terminal may store an up-to-date reference to facilitate its roaming. The information is stored in the terminal in a nonvolatile memory, just as personal telephone lists are stored. The advantage of this approach is that it obviates the need for any additional network or physical radio resources or spectrum, and it does not demand any significant cost or investment on the part of network operators, regulators, and terminal manufacturers.

The information broadcast by the global RCC would rely on data provided by operators and regulators within each country/region to a global agency like the ITU, and might include local updates on a local or national level, as well. The global radio control channel (physical or logical implementation) is expected to carry information of the following types:

- bands used for IMT-2000, including extension bands
- frequency rasters
- modulation characteristics
- identification of bands for public and private use
- guard bands
- duplex direction and spacing
- list of available services in the network
- applicable tariffs (if available)

There are regulatory as well as technical/standardization issues associated with both physical and logical RCC solutions. These issues are being addressed by ITU-R so that a consensus view could be submitted to WRC-2000 for discussion and decision.

6.3   IMT-2000 NETWORK ASPECTS

6.3.1   General

Support for evolution and/or migration of pre-IMT-2000 networks toward IMT-2000 is critical for its success. Enormous investments in second-generation technologies and network infrastructures are under way around the world. In many regions, by the time of the introduction of IMT-2000 there will have been substantial penetration of second-generation radio systems. Further, most of the regional standards for second-generation systems would have been evolved to provide more diverse services and significantly better quality of service. These (evolved) second-generation systems, representing different regional standards, will provide the platform on which IMT-2000 must be built.

As described in Section 6.2, the radio transmission technology chosen for IMT-2000 may represent a step change from the technologies deployed in the current wireless/PCS system. A number of radio transmission technologies were evaluated for IMT-2000, with a view to providing the necessary spectrum efficiency and capacity. As illustrated in Figure 6.3, the IMT-2000 networks, on the other hand, will need to evolve from the existing wireless and wireline networks because it would be prohibitively expensive to replace existing network infrastructures completely.

The standardization work on the network aspects of IMT-2000 is led by Study Group 11 (SG11) of the ITU Telecommunications Standardization Sector (ITU-T). This group is responsible for the specification of IMT-2000 signaling interfaces and also for coordination of other IMT-2000 network-related activities within ITU-T (e.g., numbering, identities, quality of service, security, management). The *IMT-2000 family-of-systems* concept adopted in the ITU-T SG11 reflects the industry view that there will be multiple IMT-2000 systems, each one belonging to a member group within the IMT-2000 family.

6.3.2   The IMT-2000 Family-of-Systems Concept

6.3.2.1   The Family Concept.   As illustrated in Figure 6.6, the IMT-2000 family of systems is a federation of IMT-2000 family members, and each system or network within a IMT-2000 family member will provide IMT-2000 capabilities to its users. These capabilities will be captured in terms of IMT-

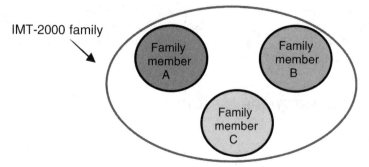

Figure 6.6 The IMT-2000 family-of-systems concept.

2000 capability sets (CS), which will be defined as CS-1, CS-2, . . . , CS-n, and will represent phased evolution of IMT-2000 standards toward increasing level of services and capabilities. Any system in the IMT-2000 family can provide service to roaming subscribers of systems belonging to other IMT-2000 family members.

Under this concept, a network belonging to an individual family member may have its own intrasystem specifications in terms of distribution of functions over physical entities (i.e., physical architecture), security mechanisms and procedures, signaling protocols, and so on. To qualify as IMT-2000 networks, however, such networks should integrate and incorporate the IMT-2000 functions into the physical entities and should support the signaling interfaces necessary to provide IMT-2000 service and network capabilities.

An IMT-2000 family member system or network is expected to have evolved from an existing wireless/wireline system, to support ITU-prescribed service and network capabilities, to support global roaming, to support ITU-defined signaling interfaces. Some additional details regarding the ITU-T specifications on these requirements on IMT-2000 networks are provided in Sections 6.3.2.2 and 6.3.2.3.

6.3.2.2   IMT-2000 Service and Network Capabilities.   According to ITU-T Recommendation Q.1701, "Framework for IMT-2000 Networks" [14], IMT-2000 functionality will be based on capability sets, which will reflect increasing levels of services and feature support as they evolve from the initial capability set 1 (CS-1) to future sets (CS-2, CS-3, etc.). The first IMT-2000 capability set (CS-1), which is captured in Recommendation Q.1701, is intended to show a significant improvement over second-generation system capabilities, its scope includes the following:

- circuit and packet bearer capability up to 144 kb/s in vehicular radio environment

- circuit and packet bearer capability up to 384 kb/s for pedestrian radio environment

- circuit and packet bearer capability up to 2048 kb/s in indoor office radio environment

- interoperability and roaming among the IMT-2000 family of systems

- service portability and support of virtual home environment

- multimedia terminals and services

- separation of call and bearer channel/connection control

- emergency and priority calls

- geographic position/location service

- user authentication and ciphering

- user–network and network–network (mutual) authentication

- lawfully authorized electronic surveillance

6.3.2.3   IMT-2000 Interfaces for Specification by the ITU.   To ensure that systems belonging to different IMT-2000 family members can interoperate to provide seamless global roaming and service delivery, the interfaces illustrated in Figure 6.7 have been identified for specification by the ITU. Definition of these interfaces requires not only the specification at the physical (transmission medium) level but also the detailed specification of protocols for exchange of signaling or control messages for support of such essential functions as call/connection control, mobility management, radio resource management, and service control. For example, definition of MT–RAN interface requires the specification not only of the physical radio interface (radio transmission technology) but also of layer 2 (data link control) and layer 3 (message delivery) protocols, as well as support for some application protocols that may be required across this interface.

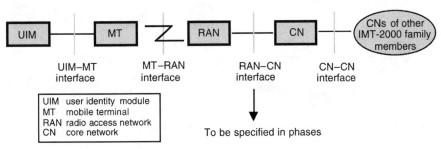

Figure 6.7 IMT-2000 family member system interfaces for specification by the ITU.

While the need for early specification of UIM–MT, MT–RAN, and CN–CN interfaces is well recognized and their specification is reaching completion as part of IMT-2000 Capability Set 1, the detailed specification of RAN–CN interface has been deferred (to CS-2 or beyond). In the initial implementations of IMT-2000 family member systems, operators may prefer to use RAN–CN interfaces based on existing wireless/PCS systems (e.g., the A interface in GSM or IS-136).

The UIM–MT interface represents the interface between a removable user identity module (similar to a SIM card in GSM) and the mobile terminal. The definition of this interface includes a secure, ISO-compliant physical specification in terms of size, contacts, electrical specifications like voltage, and basic information exchange protocols.

The CN–CN interface is a key interface for supporting global roaming across networks belonging to different family members. Though a common signaling protocol for this interface will be specified by ITU-T SG11, it is possible that in early implementations of IMT-2000 some network operators representing different IMT-2000 family member systems may prefer to design their own *interworking functions (IWFs)* based on bilateral arrangements. Figure 6.8 shows some possible scenarios for interoperation and interworking across networks belonging to different IMT-2000 family members.

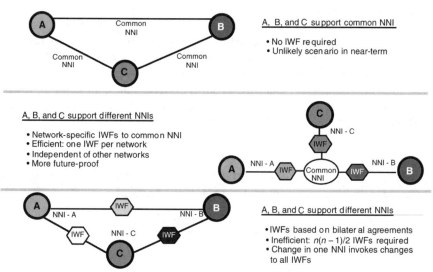

Figure 6.8 Scenarios for interworking between IMT-2000 family members: NNI, network–network interface.

6.3.3   Functional Network Architecture for IMT-2000

6.3.3.1   IMT-2000 Network Architecture: General Requirements.
IMT-2000 is expected to support a number of different radio operating environments covering indoor picocells with very high overall system capacity all the way through large outdoor terrestrial cells and satellites. A major focus of the ITU standards work on IMT-2000 systems is to maximize the commonality among the various radio interfaces, to simplify the task of building multimode mobile terminals covering more than one radio environment. A unified support of these various radio interfaces by the backbone network is therefore important. As a consequence of this requirement, it is essential that the functions that depend on radio access technology be identified and separated from functions that are not dependent on radio access technology. This separation of radio-dependent and radio-independent functions permits the IMT-2000 network to be defined, as far as possible, independently of the radio access technology.

IMT-2000 may be implemented as a stand-alone network with gateways and interworking units toward the supporting networks, in particular toward PSTN, ISDN, packet data networks (e.g., Internet), and B-ISDN (broadband ISDN). This is comparable to the current implementations of public land mobile networks, and it is also a solution in cases of fixed and radio networks that are run by different operators. However, IMT-2000 may also be integrated with the fixed networks. In this case the functions needed to support specific radio network requirements (e.g., location registration, authentication and privacy, paging and handoff) are integral to the fixed network. In such an integrated case, the base stations may be connected directly to a local exchange that can support IMT-2000 traffic by locally integrated functions and by accessing functions in other network elements. IMT-2000 systems, based on DECT, CTM (cordless telephone mobility), and PHS, are examples of such integrated implementations. Some of the RTT proposals with TDD components provide for these types of IMT-2000 implementation.

One of the key service objectives of IMT-2000 is to enable the provision of multimedia services (in circuit and packet mode operation) and Internet services (high speed packet data). Requirements for network functions must therefore take into account the support of multimedia services. IMT-2000 radio resources must be shared among circuit mode as well as packet mode services.

In addition, IMT-2000 systems should support global roaming and the virtual home environment concept: that is, the user will be provided with a comprehensive set of services and features that have the same *look and feel* regardless of whether they are accessed from the home or a visited network. The establishment of this concept recognized that service provision and network operation may be separated, allowing services to be offered by organizations not explicitly functioning as network operators. The users of these services *roam* in networks and access these services as and where commercial relationships allow it.

6.3.3.2   The ITU Three-Stage Process for Interface Specification.   The specification of an interface for IMT-2000 must provide the necessary messaging structure and parameters to support the range of IMT-2000 services and service features. Thus, the starting point for specification of interfaces in terms of detailed protocols is the set of services and service features. These are then translated into signaling requirements in terms of the following:

- a functional architecture containing required functional entities (FE) and relationships
- information flows between the functional entities and associated information elements (IE) to support various service features
- system description language (SDL) diagrams that graphically capture the information flows and functional entity actions (FEA)
- example physical architectures that identify the necessary interfaces

Detailed protocols for the interfaces identified in the last step of a stage 2 process are then developed by protocol experts using the SDL diagrams. This three-stage process for interface specification developed and deployed by the ITU is illustrated in Figure 6.9.

6.3.3.3   IMT-2000 Functional Architecture.   The broad classes of functions that need to be supported by an IMT-2000 system are illustrated in Figure 6.10, which recognizes that separation of call and connection control functions is desirable for supporting multmedia and advanced services in IMT-2000. ITU Study Group 11 has developed a detailed functional model for IMT-2000, which contains functional entities and shows the relationships between these FEs. Based on this functional model, detailed information flows have been developed which depict flow of messages between specific FEs to support individual IMT-2000 service features, and also the required information elements that need to be transported as part of these information flows. Together with the SDL diagrams, these form the stage 2 service descriptions for IMT-2000 and will be used by protocol experts to develop detailed protocol definitions across various IMT-2000 interfaces.

Figure 6.11 provides a generic reference network architecture for IMT-2000 in which the connection and call control functions are integrated (rather than separated). It illustrates one possible assignment of functional entities to physical entities, and identifies relationships between these collection of func-

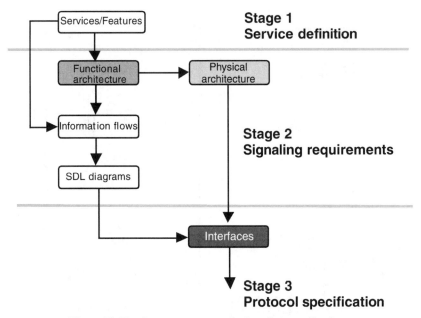

Figure 6.9 The three-stage process for interface specification.

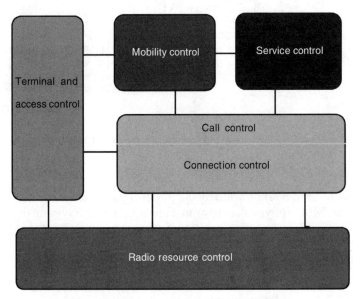

Figure 6.10 Classes of functions needed for IMT-2000.

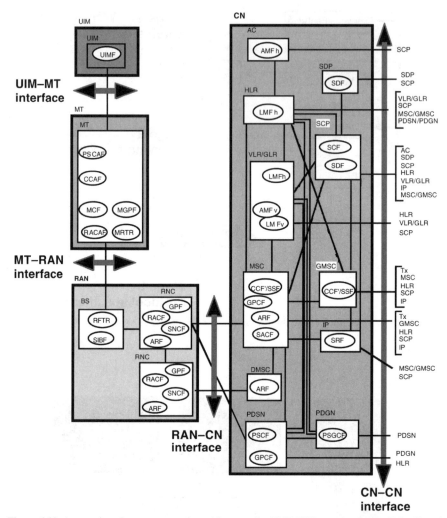

Figure 6.11 A generic reference network architecture for IMT-2000: see Appendix C for list of abbreviations.

tions that need to be supported by the physical interface specifications. For example, the MSC (physical entity) in Figure 6.11 may contain the following functional entities:

- CCF/SSF (call control function/service control function)
- GPCF (geographic position control function)
- ARF (access link relay function)
- SACF (service access control function)

Detailed descriptions of the IMT-2000 functional entities are available from the ITU [15].

6.4 SUMMARY OF REGIONAL INITIATIVES ON IMT-2000

6.4.1 Initiatives in Europe

The European Telecommunications Standards Institute, ETSI, has been working on the Universal Mobile Telecommunication System, UMTS, which will be the European standard for third generation mobile systems. UMTS will represent one of the family member groups within the IMT-2000 family of systems, and any system/network based on UMTS specifications that supports the IMT-2000 capabilities and interfaces defined by the ITU will be considered to be an IMT-2000 system/network. The focus of this activity in ETSI is the Special Mobile Group Technical Committee, which also is responsible for GSM. This activity in ETSI is supported and complemented by some of the research projects under Advanced Communication Technologies and Services (ACTS), which is funded by the European Union (EU). ETSI's objective is to develop a UMTS standard that will facilitate migration from the current GSM systems to UMTS.

ETSI has agreed on UTRA (UMTS terrestrial radio access) as the radio transmission technology for UMTS: This RTT, which utilizes W-CDMA with FDD for paired frequency bands and a hybrid W-CDMA/TDMA with TDD for unpaired frequency bands, has been submitted to the ITU-R as a candidate for IMT-2000 (see Section 6.2).

UMTS will utilize the GSM network interfaces as the basis for its network interfaces and protocols. The current GSM network capabilities, which include GPRS (general packet radio service), HSCSD (high speed circuit-switched data), and CAMEL (customized applications for mobile enhanced logic), will be enhanced in phases to support UMTS capabilities in terms of services like virtual home environment and multimedia, as well as support of higher bit rates (384 kb/s for wide-area coverage and 2 Mb/s for local-area coverage). An intermediate phase planned for GSM radio and network evolution is EDGE (enhanced data rates for GSM evolution), which will support higher data rates of 384 kb/s. Thus the IMT-2000 family member groups based on GSM will use internal interfaces evolved from current GSM systems.

6.4.2 Initiatives in North America

North America currently uses multiple radio technologies in its mobile and PCS systems. The major technologies are TDMA (IS-136), CDMA (IS-95), and GSM. Whereas the first two technologies are used both in the 800 MHz (cellular) and 1900 MHz (PCS) bands, the use of GSM is currently confined to the 1900 (PCS) band. TIA TR45 is generally responsible for the IS-136- and IS-95-based technologies, as well as for the access and network interface definitions for these technologies, the latter being covered by the ANSI-41 (TIA IS-41) specification. The GSM-related activities are generally covered by TIA TR46 in collaboration with T1P1, and both committees have close links with ETSI SMG. Unlike Europe and Japan, the United States is not planning a specific third-generation mobile system standard; rather, it is expected that the current cellular and PCS systems will evolve toward third-generation capabilities—primarily driven by market demand.

North American efforts on radio technologies for IMT-2000 are focused on a wide-band, evolved version of IS-95 CDMA and an evolved version of IS-136 TDMA. The former, referred as cdma2000, is based on direct sequence CDMA technology and is backward compatible with the current IS-95 implementations. It is compatible with the current ANSI-41 networking protocols, and its operation with GSM networks has been field-tested by some European operators. IS-136 TDMA technology is also being evolved for IMT-2000, and the evolved version is called UWC-136. Evolution of current IS-136 TDMA to UWC-136 includes development of higher packet- and circuit-switched data rates (up to 64 kb/s) in the phase 2 (target year end 1998) and data rates to 384 kb/s (wide area) and 2 Mb/s (indoor) in phase 3 (target year end 1999). UWC-136 is expected to use the EDGE concept to achieve these data rates.

As mentioned in Section 6.2, both cdma2000 and UWC-136 have been submitted to ITU-R as candidate RTTs for IMT-2000. Two additional RTTs have been submitted from the United States. The North American CDMA submitted by T1P1-ATIS is essentially the same as the UTRA proposal from ETSI and reflects the view of GSM-based PCS-1900 operators and vendors in North America. The fourth submission, called WIMS W-CDMA, is a wideband CDMA method developed by Golden Bridge Technology and submitted under the sponsorship of TIA TR46. WIMS W-CDMA is similar to the ETSI/ARIB proposals, and it aims to provide backward compatibility with ISDN/B-ISDN networks supporting multimedia services.

On the network side, TIA TR45 plans to enhance the current versions of network and access interface specifications used by both the IS-136 TDMA and IS-95 CDMA radio systems. The current ANSI-41 networking protocols are continually being enhanced to support new and advanced services, and an ad hoc group to compile IMT-2000 requirements for IS-41 has been created within TIA TR45. The access interface (A interface) specifications are also being enhanced to include ATM transport and use of AAL2 and AAL5.

6.4.3 Initiatives in Asia

Most of the third-generation wireless activities in Asia are confined to ARIB and TTC in Japan, TTA in South Korea, and CATT in China. Japan is very keen on rapid introduction of a third-generation system, hopefully in the year 2001 time frame. Japan has been a key driver and supporter of IMT-2000 standardization activities in the ITU. The current Japanese standards for wireless/PCS systems are personal digital cellular for wide-area cellular mobile and the Personal Handyphone System for local-area pedestrian applications. However, some Japanese operators have implemented North American CDMA IS-95/ANSI-41 systems.

In terms of the radio interface for IMT-2000, Japan (ARIB) has agreed on W-CDMA technology that is closely aligned with ETSI's UTRA proposal. To further align with the ETSI approach for UMTS, TTC seems to have decided to utilize the evolved GSM access and network interface implementations for IMT-2000 in Japan. TTC is working closely with ETSI SMG to ensure that Japanese requirements are fully reflected in the GSM-based UMTS network. These requirements include the use of ATM transport both in the access and core networks with AAL2 and AAL5 switching, and the possibility of IP switching in the core network. Thus, it seems that Japan is generally going to base its IMT-2000 systems on the GSM/UMTS radio and network platforms.

Mobile network operators in Japan who currently deploy, or plan to deploy, IS-95 CDMA technology with the ANSI-41 type of networking are attempting to have cdma2000 and ANSI-41 networking accepted as an alternate IMT-2000 option in Japan.

South Korea currently deploys cellular mobile networks based on the CDMA IS-95 standard utilizing ANSI-41-based networking protocols. For IMT-2000, TTA in South Korea has submitted two RTT proposals (Global CDMA I and Global CDMA II) to ITU-R. Whereas Global CDMA II is similar to the ETSI/ARIB W-CDMA proposals, Global CDMA I utilizes a lower base chip rate (0.9216 Mcps) and also calls for inter-BS synchronization.

China currently has a significant base of GSM system deployment but is now also moving toward CDMA systems. China is also becoming more active in IMT-2000 standardization work in the ITU. It has submitted a hybrid TDMA/CDMA radio transmission technology proposal to ITU-R.

6.4.4 Other Initiatives (Third-Generation Partnership Projects)

Recently the standards development organizations (SDOs) from the three major regions (Europe, North America, and Asia) decided to cooperate on developing technical specifications for some specific radio and network technologies proposed for the next generation of mobile systems. Two so-called *third-generation partnership projects* (3GPPs) have been formed to pursue this goal.

The first such partnership project is simply referred as 3GPP, and its focus is to develop technical specifications based on the wideband CDMA radio proposals from Europe (ETSI), the United States (T1P1), Japan (ARIB and TTC), and Korea (TTA), which are aligned with the European UTRA proposal, and the network infrastructure based on evolution of the GSM/MAP network. The second partnership project is known as 3GPP2, and its focus is to develop technical specifications based on the TIA TR45 cdma2000 radio and the ANSI-41 network infrastructure. The key participating standards development organizations in 3GPP2 are TIA TR45, TTC and ARIB, and TTA—representing countries that have already implemented IS-95 CDMA networks. Recently China has also become a member of 3GPP and 3GPP2. The 3GPPs are not standards development bodies, and they expect to collaborate closely with ITU-R and ITU-T in influencing standardization of IMT-2000, the international standard for the next generation of mobile systems.

6.5 Concluding Remarks

Though the ITU activities on IMT-2000 have been under way for a number of years, it is only recently that the broader commercial issues surrounding IMT-2000 and the next generation mobile systems have broken out into the wider public domain, pushing IMT-2000 to become a mainstream industry issue. The search for suitable radio technologies and associated network infrastructure to support IMT-2000 is now under way, with significant participation from key industry segments. Consensus is now emerging for ITU-R TG8/1 to provide a single, flexible standard for IMT-2000 radio interface that will support multiple radio technologies. These will perhaps include two harmonized CDMA modes, two TDMA modes, and one or two hybrid modes with as much commonality as possible in the RF parameters to facilitate the design of multi mode terminals.

Besides the difficult task of specifying suitable radio transmission technology for IMT-2000, currently under way in ITU-R, a significant component of this activity is related to the development of signaling protocols at the radio, access, and network interfaces. In addition to supporting a broad category of services (voice, packet data, and multimedia) to provide an efficient yet flexible network infrastructure to accommodate diverse radio environments (indoor, urban, suburban, rural and satellite) and potential radio technologies (CDMA, TDMA, hybrid) that IMT-2000 may entail.

Also gaining momentum are standardization activities in other related areas, such as service definitions, network operations (numbering, identities, performance, management), security, charging and accounting, encoding and multiplexing, and network and service interworking. In all cases, the underlying objective is to complete the IMT-2000 specifications around year 2000.

6.6   REFERENCES

[01] F. Liete et al., "Regulatory Considerations Relating to IMT-2000," *Personal Communications Magazine*, Vol. 4, No. 4, August 1997.

[02] R. D. Carsello et al., "IMT-2000 Standards: Radio Aspects," *Personal Communications Magazine*, Vol. 4, No. 4, August 1997.

[03] Universal Mobile Telecommunications Forum, "IMT-2000 Spectrum Requirements," Special Ad Hoc Group Report, UMTS Forum, London, 1997.

[04] ITU-R Recommendation M. 1036 (revised), "Spectrum Considerations for Implementation of International Mobile Telecommunications 2000 (IMT-2000) in the Bands 1885–2025 MHz and 2110–2200 MHz," International Telecommunication Union, Geneva, 1998.

[05] ITU-R Recommendation M. 1225, "Guidelines for Evaluation of Radio Transmission Technologies for IMT-2000," International Telecommunication Union, Geneva, 1997.

[06] ITU Web site for IMT-2000 Radio Transmission Technology Proposals: *http://www.itu.int/imt/2-radio-dev/rtt-proposals/index.html*

[07] K. Jamal, "The UTRA Concept for IMT-2000," IMT-2000 Evaluation Workshop, Seoul, South Korea, July 14–15, 1998.

[08] Association of Radio Industries and Business, "Japan's Proposal for Candidate Radio Transmission Technology on IMT-2000 W-CDMA," IMT-2000 Evaluation Workshop, Seoul, South Korea, July 14–15, 1998.

[09] N. Ramesh, "cdma2000 Radio Transmission Technology (RTT) for IMT-2000," IMT-2000 Evaluation Workshop, Seoul, South Korea, July 14–15, 1998.

[10] P. Meche, "UWC-136 RTT Overview," IMT-2000 Evaluation Workshop, Seoul, South Korea, July 14–15, 1998.

[11] ITU-R Delayed Contribution No.8-1/85-E, "Discussion Document on the Global Radio Control Channel", ITU-R Task Group 8/1 Meeting, Geneva, April/May 1998.

[12] R. Pandya et al., "IMT-2000 Standards: Network Aspects," *Personal Communications Magazine,* Vol. 4, No. 4, August 1997.

[13] ITU-T Recommendation Q.1701, "Framework for IMT-2000 Networks," International Telecommunication Union, Geneva, 1999.

[14] ITU-T Recommendation Q.1711, "Network Functional Model for IMT-2000," International Telecommunication Union, Geneva, 1999.

[15] R. Pandya, "Network Architecture, Interfaces and Signaling Protocols for IMT-2000," *Nikkei Business Review* (in Japanese), July 1998.

[16] J. S. Dasilva et al., "European Third-Generation Mobile Systems," *Communications Magazine,* Vol. 34, No. 10, October 1996.

CHAPTER 7

GLOBAL MOBILE
SATELLITE SYSTEMS

7.1 INTRODUCTION

To make a satellite phone call today from a location that does not offer terrestrial wireline or wireless coverage requires the use of a large, costly terminal, and entails very high per-minute charges. Further, the quality of service is
relatively poor because of annoying echoes, large transmission delays, and
overtalk associated with satellite communications using geostationary satellites. The next generation of satellite communication systems will use advances
in satellite systems, wireless technology, and miniaturization to provide global
mobile satellite services that will make calls between any two locations on
earth much easier, much more affordable, and much more user-friendly.

Even in the year 2000, terrestrial cellular coverage is available to less than
60% of the world's population and on only about 15% of the earth's total surface. Rural areas, regions that are sparsely populated in developed countries,
and large parts of the developing world are destined to be underserved or to
remain out of the reach of terrestrial mobile services altogether. Thus, in many
parts of the world, the demand for communications mobility can be met effectively only through global mobile satellite services. Handheld satellite
phones are therefore forecast as the emerging mobile communication frontier,
with growth that could parallel recent growth in the cellular mobile industry.

Annual revenue forecasts for mobile satellite services range in the order
of $12 billion by year 2004, and subscriber forecasts are in the range of $35 million by year 2010. As indicated in Chapter 6, on next-generation mobile systems, global mobile satellite systems are a key component of IMT-2000, and
the frequency spectrum identified for IMT-2000 includes an allocation for mobile satellite services. The emerging next-generation satellite mobile systems
are generally referred as GMPCS, for global mobile personal communication
by satellites.

215

Until now, communications satellites have operated using a geostationary orbit (GEO) lying about 36,000 km above the earth's surface. From this orbit the satellite appears to be stationary (fixed) above a specific location from earth, thereby ensuring continuous, uninterrupted coverage to that location. The primary role of a geostationary communications satellite is to act as a wireless repeater station in space that operates in a broadcast mode and provides a microwave link between two remote locations on earth. The key components of a communications satellite include various transponders, transceivers, and antennas that are tuned to the allocated frequency channels.

Although the geostationary satellites have a large footprint, so that the entire surface of the earth can be covered by a few such satellites, their high altitude leads to very long round-trip signal delays and resultant degradation in service quality. To support a wide range of services and provide superior service quality comparable to that available from terrestrial wireless and wireline networks, constellations of satellites operating in low earth orbits (LEO) or medium earth orbits (MEO) are considered more suitable. Table 7.1 provides some key characteristics of LEO, MEO, and GEO satellites.

A number of global mobile satellite systems are in various stages of planning and deployment, with the first global mobile satellite service initiated in 1998. The four such systems that are in advanced stages of planning and/or early implementation are Iridium, Globalstar, ICO, and Teledesic. While the primary target applications for the first three systems are voice, fax, and messaging services for mobile communication subscribers, Teledesic system's target application is high speed data and multimedia services using satellites. Except for the ICO system, which deploys a constellation of MEO satellites, these mobile satellite systems use various size constellations of LEO satellites.

Table 7.1 Characteristics of Different Satellite Configurations

Characteristic	Satellite Type		
	GEO	**MEO**	**LEO**
Altitude range	36,000 km	10,000–20,000 km	500–2000 km
Satellite visibility	24 hours	2–4 hours	10–20 minutes
Round-trip delay	500 ms	40–80 ms	5–10 ms
Satellite lifetime	20–30 years	10–15 years	4–8 years
Satellite constellation cost	Low	Medium	High

To provide global coverage for mobile subscribers, LEO systems need to deploy a large number of satellites; they must either support intersatellite links (Iridium, Teledesic) or use a large number of ground stations (Globalstar). These factors, combined with the requirement for more frequent replacement of LEO satellites, may lead to overall higher costs for LEO systems than for MEO and GEO satellite systems. LEO systems also face additional technical challenges because of frequent switching of phone calls from one satellite to another (handoff) and potential susceptibility to shadowing (loss of signal due to shadows cast by buildings, etc.) associated with low orbits. Sections 7.2 to 7.5 provide brief descriptions of these four global mobile satellite systems.

7.2   THE IRIDIUM SYSTEM

The Iridium system is not proposed to be a replacement for existing terrestrial cellular systems, but rather as an extension of existing wireless systems to provide mobile services to remote and sparsely populated areas that are not covered by terrestrial cellular services. It provides more capacity (large number of channels) and better quality of service (shorter transmission delays) to areas that currently receive mobile services from geostationary satellites (e.g., maritime mobile services). It can also provide emergency service in the event that terrestrial cellular services are disabled in disaster situations (earthquakes, fires, floods, etc.).

The concept of using a constellation of low earth orbit satellites to provide global telecommunications services to mobile users was originated by Motorola in 1987. Because the initial proposal called for 77 satellites in the constellation, the system was called Iridium after the element, which has 77 electrons in its orbit. Later studies indicated that only 66 satellites would be adequate to provide the targeted services and performance. The launch of the 66 LEOs has been completed and the service is now in operation. The 66 satellites are grouped in six orbital planes; there are 11 active satellites in each plane with uniform nominal spacing of 32.7°. The satellites have circular orbits at an altitude of 783 km, and for each plane an in-orbit spare satellite is provided.

The satellites travel up one side of the earth, crossing over near the north pole, then traveling down the other side. With 11 satellites in each of the 6 planes, both sides of the earth are covered on a continuous basis by the 66 active satellites. As illustrated in Figure 7.1, satellites in one plane are placed to travel out of phase with those in the adjacent planes. Except for the first and last planes, which are counterrotating where they are adjacent, all remaining planes are corotating. The distance between corotating planes is 31.6°, and the distance between the counterrotating planes is 22°. The reduced separation between counterrotating planes is needed to compensate for the reduced cover-

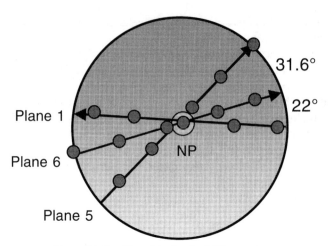

Figure 7.1 Satellite orbits for the Iridium system.

age provided by satellites on counterrotating planes. In the Iridium system, each satellite is equipped with four two-way communication links (intersatellite links, or ISLs), one each with its neighbors in the same plane and with those in the adjacent planes.

Each Iridium satellite uses a 48-beam antenna pattern, and each beam, which has a minimum diameter of 600 km, can be individually switched. For example, only about two-thirds of the beams will be active at any given time because some of the beams will be switched off when the satellites are in the vicinity of the poles, where beam patterns tend to overlap, or when the satellites are over countries or regions in which, Iridium does not have regulatory arrangements to operate. The switching of beams is referred as *cell management*. In a LEO-based system like Iridium, the beams are equivalent to cells associated with terrestrial mobile systems. However, in case of the Iridium system, it is the beams that move rapidly relative to the subscriber, who is considered to be stationary with respect to the satellite. Thus, switching of beams or cell management to provide continuity of an existing call is equivalent to handoff in terrestrial cellular mobile systems. This requirement for cell management is, of course, an additional complexity associated with LEO-based systems compared with MEO or GEO systems.

The Iridium system supports links of three types: up- and downlinks from the space vehicle (SV) to the gateway (GW) [or to the telemetry, tracking, and control (TT&C) center], using the Ka band; up-and downlinks between the SV and the Iridium subscriber unit (ISU), using the L band; and two-way intersatellite links between the SVs using the Ka band. The parameters associated with these links are summarized in Table 7.2.

Table 7.2 Technical Parameters for Satellite Links Supported by Iridium

Link Parameter	SV ⇌ ISU	SV ⇌ GW	SV ⇌ SV
Frequencies, GHz	1.616–1.625 (UL) 1.161–1.625 (DL)	27.5–30.0 (UL) 18.8–20.2 (DL)	22.55–23.55
Center frequency	1.62125 GHz	29.4 GHz (UL) 20.0 GHz (DL)	23.28 GHz
Multiple access	TDMA/FDMA	TDMA/FDMA	TDMA/FDMA
Modulation	QPSK	QPSK	QPSK
Polarization	Right-hand circular	Right-hand circular	Vertical
Total bandwidth	10.5 MHz	100 MHz	200 MHz
Channel bandwidth	31.5 kHz	4.375 MHz	17.5 MHz
Channel spacing	41.67 kHz	7.5 MHz	25 MHz
Coded data rate	0.05 Mb/s (UL) 0.05 Mb/s (DL)	6.25 Mb/s (UL) 6.25 Mb/s (DL)	25.0 Mb/s
Final amplifier output power	0.1–3.5 W per carrier (burst)	0.1–1.0 W per channel	3.4 W per carrier (burst)

A high-level network architecture for the Iridium system is illustrated in Figure 7.2. Connections between the Iridium satellites and the PSTN are provided via gateways, generally colocated with earth stations (ES). The intersatellite links between the neighboring satellites provide flexibility for location of earth stations and remove the requirement that an ES be continuously available within the satellite footprint. The appropriate ISLs can be used to route each call to the ES closest to the origination or destination of the call. An ISU-initiated call can be routed within the satellite network to another ISU located anywhere on earth, or it can be connected to the public switched network through an earth station for routing and delivery through the PSTN.

The ISU will be a dual-mode mobile station that will support both satellite and terrestrial mobile network interface standards. Thus in addition to the Iridium satellites, it will provide access to a public land mobile network operating in one of the PLMN frequency bands allocated for cellular and personal communication services (PCS). The call processing architecture for the Iridium system is based on GSM, and each ES incorporates the GSM MSC functions as well as functions associated with such databases as EIR, VLR, and HLR. The ES also supports functions that are specific to the satellite operation. These satellite network related functions include control of feeder links, ES management, and message control. Example call flows for a PSTN-to-ISU call in the Iridium system and for a call between two ISUs are

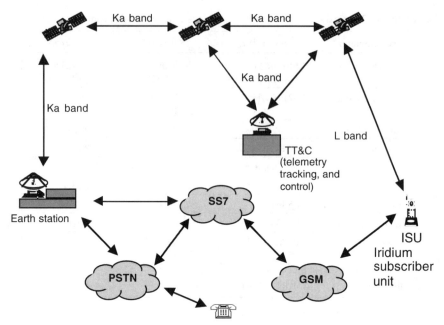

Figure 7.2 High level network architecture for the Iridium system.

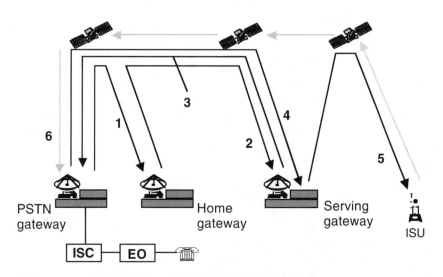

1 PSTN GW sends ISU's MSISDN to home GW.
2 Home GW querries serving GW for ISU location information.
3 Serving GW returns location information to PSTNN GW.
4 PSTN GW routes call to serving GW.
5 Serving GW alerts terminating ISU.
6 Voice path after ISU answers.

Figure 7.3 Call flow example for a PSTN-to-ISU call in Iridium.

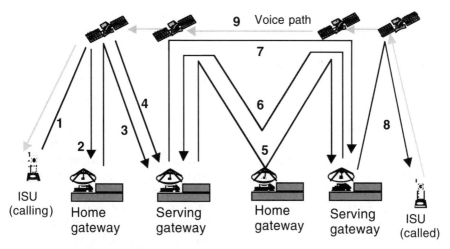

1 ISU acquires SV.
2 Access to home GW (calling) for authentication.
3 Service transferred to servicing GW (calling) after authentication.
4 Setup request to servicing GW (calling).
5 Location querry to servicing GW (called).
6 Location information to servicing GW (calling).
7 Route call to servicing GW (called).
8 Ring alert ISU (called).
9 Establish voice path on answer.

Figure 7.4 Call flow example for an ISU-to-ISU call in Iridium.

shown in Figures 7.3 and 7.4, respectively. These call flows are quite similar to those in GSM networks.

7.3   THE GLOBALSTAR SYSTEM

Globalstar is a global mobile satellite system based on a constellation of 48 LEO satellites. The launch of these satellite is in progress and initial service offering is scheduled for late 1999. Unlike the Iridium system, Globalstar system does not use intersatellite links but rather depends on a large number of interconnected earth stations or gateways for efficient call routing and delivery over the terrestrial network. It is designed to complement the terrestrial cellular mobile networks to provide telephony and messaging services to subscribers in locations that are not covered, or inadequately covered, by conventional wireline or wireless networks.

Globalstar's constellation of 48 LEO satellites is designed to orbit at an altitude of 1414 km above earth's surface in eight orbital plans inclined at 52°. With each plane to be occupied by six satellites with a provision for one in-orbit spare satellite in each plane. The nominal weight of each satellite is

450 kg, with a deployed span of 7 meters and working life of 7.5 years. Since Globalstar satellites do not employ intersatellite communication, they essentially provide only transponder functions, making their design and operation less complex and perhaps more reliable. Each satellite supports a 16-beam antenna pattern with an average beam diameter of 2250 km. To mitigate blocking and shadowing, Globalstar will deploy path diversity, whereby multiple satellites may be used to complete a call.

In the absence of intersatellite links, the Globalstar system *makes* maximum use of the international terrestrial networks (wireline and wireless). Calls from a subscriber are routed via a satellite to the nearest earth station/gateway, and from there they will be routed over the existing terrestrial network. To provide the interface between the ground segment (terrestrial networks) and the space segment (Globalstar satellites) Globalstar design deploys 100 or more gateway stations distributed around the world with each station equipped with three to five antennas that can track the trajectories of the satellites. A Globalstar gateway is designed to serve an area 3000 km in diameter and will be designed to take into account the technical and administrative requirements of the coverage area. These requirements may include such factors as coverage, quality of service, and satellite visibility, as well as regulatory and contractual factors associated with national boundaries.

Globalstar uses two types of communication links: service links in the L/S band for communication between the terminals and the space vehicle, and the gateway links in the C band for communication between the earth stations and the space vehicle. The actual frequencies used for these links are as follows:

Service Links (SV ⇌ Terminal)

Terminal to SV	1610.0–1625.5 MHz (L band)
SV to Terminal	2483.5–2500.0 MHz (S band)

Gateway Links (SV ⇌ GW)

GW to SV	5091.0–5250.0 MHz (C band)
SV to GW	6875.0–7055.0 MHz (C band)

Globalstar uses CDMA multiple access technology for the service links and FDMA/FDD for the gateway links with QPSK modulation. CDMA provides increased capacity through its frequency reuse, voice activity detection, and spectrum-sharing capabilities, and better performance through its support for multipath diversity. The network architecture for the Globalstar mobile satellite system (Figure 7.5) is very similar to the Iridium system except that Globalstar does not support ISLs.

Globalstar will support a range of terminals including fixed (public call box, PABX), mobile (automobiles), and handheld terminals. The following

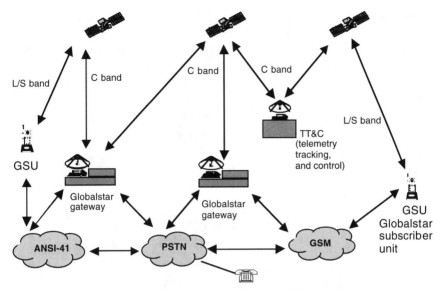

Figure 7.5 Network architecture for the Globalstar system.

four versions of handheld terminals, with maximum transmit power capability of 2 W, will be used by Globalstar:

- single-mode: Globalstar
- dual-mode: Globalstar/GSM
- dual-mode: Globalstar/CDMA cellular
- triple-mode: Globalstar/CDMA cellular/AMPS cellular

The subscriber can therefore choose the type of handheld terminal based on the mobility and roaming requirements.

The Globalstar system will provide interconnection to PSTN and PLMNs through 100 to 200 gateways distributed around the world. The GW architecture will incorporate MSC and database functions for GSM networks as well as ANSI-41 networks, thus offering support to subscribers from different regional cellular systems. The gateway architecture is flexible enough to support multiple GSM MSCs within a single GW, which may also have an ANSI-41 MSC. A single GSM MSC can also be shared between multiple GWs.

Calls in the Globalstar system utilize the satellite links only when connections cannot be made through the terrestrial wireline or wireless networks. The calls are connected through the regional gateways so that the local service providers can share the revenue and the administrations can ensure regulatory compliance. Satellite operation control centers (SOCCs) tracks and controls the satellites, and ground operation control centers (GOCCs) dynamically allocates capacity between adjoining regions, coordinate information from SOCCs, and collect billing information. Figure 7.6. shows an example of a call setup in the Globalstar network from a Globalstar/AMPS terminal outside the coverage of terrestrial facilities to a Globalstar/GSM terminal roaming in a GSM network.

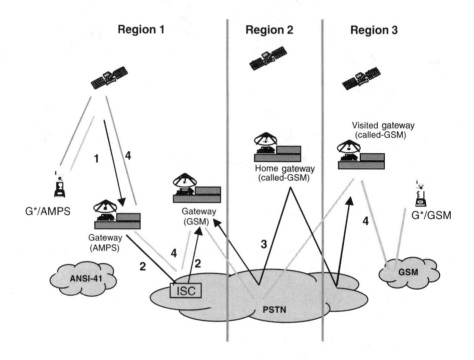

1   In the absence of terrestrial coverage, call request routed to GW via satellite.
2   Call routed via PSTN ISC to nearest GW with GSM MSC functions.
3   GW (GSM) obtains subscriber location information from called party's visited GW (via home GW).
4   Globalstar GW (GSM) in Region 1 establishes voice path after called party answer via PSTN and visited GSM.

Figure 7.6 Example of a call setup in the Globalstar system: solid lines, signaling; hatched lines, voice path.

7.4   THE ICO SYSTEM

The ICO is a medium earth orbit (MEO) mobile satellite system, which is designed primarily to provide services to handheld phones. ICO will use TDMA as the radio transmission technology. ICO has submitted this proposal to the ITU-R evaluation/selection process as a potential candidate RTT for the satellite component of IMT-2000—the third-generation mobile telecommunication system being specified by the ITU. The ICO system is planned to go in service in August 2000. The system is designed to offer digital voice, data, facsimile, and short targeted messaging services to its subscribers. ICO's primary target customers are users from the existing terrestrial cellular systems who expect to travel to locations in which coverage is unavailable or inadequate. Other customer groups potentially served by ICO include road transport, maritime, and aeronautical users, as well as users of semifixed terminals in rural areas and developing countries, where conventional terrestrial wireline or wireless facilities are not cost-effective. The ICO system design is intended to integrate mobile satellite communications capability with the public land mobile networks like GSM, D-AMPS, and PDC and their PCS variants.

ICO system is designed to use a constellation of 10 MEO satellites in intermediate circular orbit (ICO), at an altitude of 10,355 km above the earth's surface. The nominal weight of these satellites at launch, is less than 2000 kg. The satellites, with an expected life of 12 years, are arranged in two planes with five satellites (and one spare) in each plane: orbital planes inclined at 45° relative to the equator. Each satellite has antennas to provide 163 transmit and receive service link beams. The orbital configuration provides coverage of earth's entire surface at all times and ensures significant overlap so that two or more satellites are visible to the user and the satellite access node (SAN) at any time. Further, at least one of the satellites appear at a high elevation angle, thereby minimizing the probability of blocking due to shadowing effects.

The ground segment in the ICO system, which will link the ICO satellites to the terrestrial networks, will consist of 12 interconnected SANs located in various parts of the world. Each SAN consists of earth stations with multiple antennas for communication with the satellites, switching equipment, and databases to facilitate interconnection with public telephone, data, and mobile networks. The interconnection to the public networks is through appropriate gateways. Whereas each SAN supports VLR functions, the HLR function can reside in one (or more) of the SANs. A SAN tracks the satellites within its sight and will direct communication traffic to the satellite, which can provide reliable, uninterrupted link for a given call, in terms of angle of elevation and duration of satellite visibility. SANs also have the capability to execute handoffs from an area covered by one satellite to another satellite's coverage. Such handoffs are expected to be very infrequent in ICO's MEO-based system. Besides the SANs, the ICO system deploys TT&C stations connected to a satellite control center (SCC) for monitoring and controlling the satellites, as well as one or more network control centers (NCC) for overall management and control of the ICO system. The TT&C functions are associated with 6 of the 12 interconnected SANs. Overall configuration for the ICO system is illustrated in Figure 7.7.

As indicated in Figure 7.7, the S-band frequencies will be used for communication between the satellite and the ICO user terminals. The frequencies selected for this purpose (2170–2200 MHz for uplink and 1980–2010 MHz for downlink) are consistent with the WARC-92 allocation of frequencies for the satellite component of IMT-2000. The links between the satellite and SANs (feeder links) will utilize the C band. The C band frequencies ICO expects to use

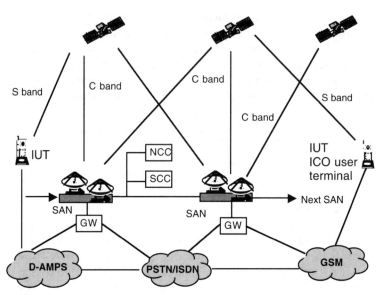

Figure 7.7 Overall configuration for the ICO system.

for the feeder links are 5100–5250 MHz for uplink and 6935–7075 MHz for downlink. These bands form part of a pair of new allocations made by WARC-95 for feeder links to nongeostationary satellites used for mobile satellite services.

ICO user terminals will be predominantly handheld, pocket-sized telephone terminals capable of dual-mode operation with satellite and cellular/PCS systems based on GSM as well as North American and Japanese digital cellular/PCS standards. As satellite components of IMT-2000, ICO terminals in the future will also have capabilities to operate in both the satellite and terrestrial components of IMT-2000. These battery-operated terminals will weigh between 180 and 250 grams, with talk time and standby times of 4 to 6 hours and 80 hours, respectively.

The following considerations and/or principles are employed in the ICO system for call routing and delivery:

- SANs are linked by the inter-SAN network for signaling and traffic.
- Gateways provide the only terrestrial access between the PSTN and the SANs.
- A gateway is connected to at least one SAN.
- An ICO subscriber is always associated with an HLR, generally located at a SAN.
- A roaming ICO subscriber is also associated with a VLR located at a SAN.

- An ICO user terminal uses the nearest SAN for call origination/ termination.

- A PSTN or cellular terminal uses the nearest SAN for call origination/termination.

- Calls between two ICO terminals are routed via the inter-SAN network.

Figure 7.8 diagrams an example call flow for a call originating in a terrestrial cellular network and terminating on an ICO user terminal outside the range of a terrestrial network.

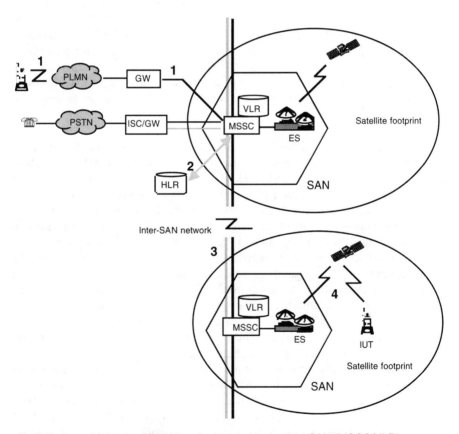

1 Call request from the PLMN terminal routed to nearest SAN (MSCC/VLR).
2 MSCC/VLR obtains location information (serving SAN for called party) from HLR.
3 Call routed to serving SAN (MSCC/VLR) over the inter-SAN network.
4 Call completed to called IUT over the satellite segment.

Figure 7.8 An example of a call setup in the ICO system.

7.5   THE TELEDESIC SYSTEM

Currently the high bandwidth, high quality fiber connectivity needed to support Internet access, computer networking, video conferencing, and so on is restricted to major commercial and population centers. Outside these application areas, such facilities are either too expensive or simply not available. The aim of the Teledesic network is to extend the existing terrestrial, fiber-based infrastructure to provide advanced information and communication services anywhere on earth. Whereas the target application for Iridium, Globalstar, and ICO is voice, with support of low bit rate data for facsimile and messaging for mobile subscribers, the primary target application for the Teledesic system is to provide worldwide, seamless, fiberlike connectivity to support multimedia, video, and high bit rate data services. *In a strict sense, Teledesic does not fall in the category of global mobile satellite systems or GMPCS because its focus is not on worldwide terminal mobility, but rather on providing the so-called Internet in the sky function.* The planned target for Teledesic service availability is end of year 2002. Rather than individual end users, primary customers for the Teledesic system will be service providers in countries around the world wishing to extend their network capabilities in terms of geographic scope and the range of services, and also multinational corporations needing to extend the capabilities of their enterprise networks.

The design of the Teledesic system has not been finalized. According to the original plans, the Teledesic satellite segment was to use 840 LEO satellites in 21 planes at altitudes of 700 km. The Teledesic system now intends to deploy only 288 active LEO satellites placed in 12 planes (24 satellites per plane) at altitudes around 1350 km. Each satellite in the Teledesic constellation will have connections to eight of its neighboring satellites through intersatellite links operating in a connectionless packet mode, with each satellite in this interconnected mesh network providing necessary switching functions. The Teledesic network is designed for dual-satellite visibility with at least one insight satellite at a minimum elevation of 40°. This high elevation angle ensures an unobstructed and omnidirectional view of the sky from most building tops where Teledesic terminals may be located. Besides eliminating shadowing effects from neighboring buildings and terrain, the high elevation angle greatly reduces the fading effects of rain at high frequencies.

The frequency spectrum for the intersatellite links will be in 60 GHz band. The terrestrial networks (PSTN, PLMN, private networks) will communicate with the Teledesic satellites via gateways and earth stations using the Ka-band frequencies (28.6–29.1 GHz for uplink and 18.8–19.3 GHz for downlink). Alternatively, Teledesic customers can also use a specially designed Teledesic terminal to communicate directly with the satellite using Ka-band frequencies. The terminal can be mounted flat on a rooftop and connected inside a building with the customer's computer network. Figure 7.9 presents a high level

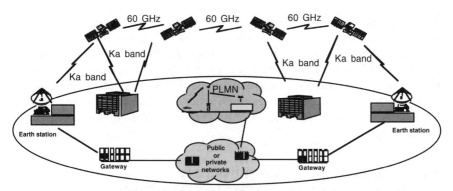

Figure 7.9 High level view of the Teledesic system.

view of the Teledesic system. The transmission mode within the Teledesic network is based on streams of short, fixed-length packets of 64 octets (512 bits), which is similar to the asynchronous transfer mode (ATM) transport deployed in most terrestrial fiber networks. The intersatellite links have a capacity of 155.52 Mb/s each, and the bit rates for links between the satellites and the gateways range from 155.52 Mb/s up to 1244.16 Mb/s, in increments of 155.52 Mb/s.

To minimize the number of handoffs due to the motion of the satellites and the earth's rotation, Teledesic will use steerable antennas and regional resource mappings. The entire earth's surface will be divided into a hierarchy of cells and supercells, with 9 cells per supercell. A Teledesic satellite's footprint will cover a maximum of 64 supercells (576 cells), corresponding to one supercell per beam. The frequencies and time slots (channel resources) will be associated with each cell and managed by the serving satellite. Thus, more than one satellite may serve the mobile terminal during the duration of the call, but the channel resources assigned to the call will remain unchanged. The use of this cells-and-supercell approach combined with onboard databases on the satellites also allows Teledesic to customize the service for certain geographic areas/countries and to minimize interference to are from specific areas.

The multiple-access method used in the Teledesic system is a combination of TDMA, FDMA, SDMA (space division multiple access), and ATDMA (asynchronous TDMA). Within a supercell, TDMA is deployed between cells, where a beam scans each cell in the supercell with a scan cycle of 23.11 ms and a scanning beam capacity of 1440 channels of 16 kb/s. Since all supercells receive transmissions from the satellite at the same time, SDMA is used between cells scanned simultaneously in adjacent supercells. Within the time slots assigned to individual cells (in a supercell), mobile terminals use FDMA for uplink and ATDMA for downlink. On the downlink, the header in the ATDMA packets is used to route packets to a specific terminal. All available frequencies or a fraction thereof can be assigned to a terminal for uplink transmission for entire duration of the call. Further, the TDMA operation between cells implies that only one of the nine cells (in a supercell) can use all available frequencies at a time, corresponding to a frequency reuse pattern of 9. Since the Teledesic systems plans to have 20,000 supercells, the resulting global frequency reuse factor works out to be 2222.

7.6 SUMMARY TABLES FOR THE IRIDIUM, GLOBALSTAR, ICO, AND TELEDESIC SYSTEMS

Tables 7.3 to 7.5 provide a comparative view of the Iridium, Globalstar, ICO, and Teledesic global mobile satellite systems, as well as their general characteristics and some key technical parameters.

Table 7.3 General Characteristics of Iridium, Globalstar, ICO, and Teledesic Systems

Characteristic	Iridium	Globalstar	ICO	Teledesic[a]
Primary backers	Motorola Raytheon	Loral Qualcomm	Inmarsat Hughes Space	Craig McCaw and Bill Gates
System type	LEO	LEO	MEO	LEO
Target services	Voice, fax, low speed data	Voice, fax, low speed data	Voice, fax, low speed data	Multimedia, high speed data
Voice channels	2.4/4.8 kb/s	2.4/4.8/9.6 kb/s	4.8 kb/s	16 kb/s
Data throughput	2.4 kb/s	7.2 kb/s	2.4 kb/s	16-2048 kb/s
Modulation	QPSK	QPSK	QPSK	—
Voice channels per satellite	1100	2000–3000	4500	100,000, 16 kb/s each
E.164 global numbering code	881-6 and 881-7	881-8 and 881-9	881-0 and 881-1	882-x (pending)
E.212 global identity code	901-03	901-04	901-01	N/A
Terminal types	Dual mode, mobile, handheld	Dual mode, mobile, handheld	Dual mode, mobile, handheld	Single mode, nomadic, portable
FCC licensed	Yes	Yes	No	Yes
Service launch	1998	1999	2000	2002

[a] System design not yet frozen—some characteristics may be revised.

Table 7.4 Space Segment Characteristics of Iridium, Globalstar, ICO, and Teledesic Systems

Characteristic	Iridium	Globalstar	ICO	Teledesic[a]
Number of satellites	66 + 6 spares	48 + 8 spares	10 + 2 spares	288 + 12 spares
System type	LEO	LEO	MEO	LEO
Number of planes	6	8	2	12
Orbit altitude	783 km	1 414 km	10,355 km	1 350 km
Inclination	86.4°	52°	45°	98.16°
Minimum elevation	8.2°	10°	10°	40°
Footprint	4700 km	5850 km	12,900 km	1412 km
Satellite visibility	11.1 min	16.4 min	115.6 min	3.5 min
Round-trip delay	~ 10 ms	~ 10 ms	~ 200 ms	~ 10 ms
Intersatellite links	Yes	No	No	Yes
Gateways	12	100–210	12	—
Satellite antenna type	Fixed (moving cells)	Fixed (moving cells)	Fixed (moving cells)	Steerable (earth-fixed cells)
Satellite lifetime	5–8 years	7.5 years	15 years	10 years
Satellite mass	700 kg	450 kg	1925 kg	770 kg
Satellite output power	1400 W	1000 W	2500 W	—

[a] System design not yet frozen—some characteristics may be revised.

Table 7.5 Link and Communications Parameters for Iridium, Globalstar, ICO, and Teledesic Systems

Characteristic	Iridium	Globalstar	ICO	Teledesic[a]
Frequencies; SV to terminal	1616–1625.5 MHz (L band)	2483.5–2500 MHz (S band)	1980–2010 MHz (S band)	18.8–19.3 GHz (Ka band)
Terminal to SV	1616–1625.5 MHz (L band)	1610–1625.5 MHz (L band)	2170–2200 MHz (S band)	28.6–29.1 GHz (Ka band)
SV to GW	18.8–20.2 GHz (Ka band)	6.875–7.055 GHz (C band)	6.935–7.075 GHz (C band)	18.8–19.3 GHz (Ka band)
GW to SV	27.5–37.0 GHz (Ka band)	5.091–5.250 GHz (C band)	5.1–5.25 GHz (C band)	28.6–29.1 GHz (Ka band)
SV to SV	22.55–23.55 GHz (Ka band)	N/A	N/A	60 GHz band
Multiple access	TDMA/FDMA/ TDD	CDMA/FDMA/ FDD	TDMA/FDMA/ FDD	TDMA/SDMA/ FDMA/ATDMA
Beams/satellite	48	16	163	
Reuse pattern	12 cells/cluster	1 cell/cluster	4 cells/cluster	9 cells/cluster

[a] System design not yet frozen—some characteristics may be revised.

7.7   CONCLUDING REMARKS

Besides the Iridium, Globalstar, ICO, and Teledesic systems, which intend to provide global coverage, a number of regional mobile satellite systems are planned. These regional systems intend to use one or more geostationary satellites to provide services in specific regions. Examples of such regional mobile satellite systems include the Thuraya system in the Middle East and the APTM (Asia–Pacific Mobile Telecommunications) system in Southeast Asia.

Similarly a number of satellite broadband networks besides Teledesic, which intend to provide satellite–based network infrastructure for video conferencing and high bit rate data services are being planned. The major networks in this category include Cyberstar (Loral Corporation), Astrolink (Lockheed), Spaceway (GM–Hughes), and Skybridge (Alcatel and Loral). Most of these systems plan to launch their services in the 2000–2002 time frame, using Ka-band or Ku-band frequencies.

Besides the technical aspects for the design and operation of GMPCS systems, telecommunications specialists attempting to provide effective and efficient delivery of global mobile satellite services face some special regulatory hurdles. Though these systems are global in nature, they still need to satisfy appropriate regulations in each country where they intend to provide service. Some of the regulatory aspects associated with GMPCS are as follows:

- frequency allocation and coordination
- spectrum sharing (conflicts in spectrum usage)
- licensing and authorization
- network interconnection
- mobile station E.164 numbers and E.212 identities
- mobile terminals—type approvals
- protection of sovereign rights and interests

ITU is playing a key role in facilitating consultation and cooperation efforts among its members (national administrations and industry) to ensure adoption of worldwide coordination regimes and licensing policies. To this end, ITU held a World Telecommunications Policy Forum in Geneva (October 21–23, 1996) to foster an exchange of views and information on policy and regulatory issues raised by the introduction of GMPCS systems.

7.8   REFERENCES

[01] F. Leite, "Global Mobile Personal Communication Systems: The Technical Challenge," *ITU News,* December 1996.

[02] *www.ee.surrey.ac.uk/Personal/Wood/Constellations/*

[03] T. Logsdon, *Mobile Communication Satellites—Theory and Applications* McGraw-Hill, New York, 1995.

[04] F. Abrishamkar and Z. Siveski, "PCS Global Mobile Satellites," *Communications Magazine,* Vol. 34, No. 9, September 1996.

[05] Iridium Network Overview, Briefing for ITU-T Study Group 2, Geneva, September 19–29, 1995.

[06] J.-B. Lagarde, "Mobile Satellite Communication Services," *Alcatel Telecommunications Review,* Second Quarter 1997.

[07] Globalstar Call Flow Diagrams, briefing to ITU-T Study Group 2, March 25, 1996.

[08] ICO system description, ICO Global Communications, London, February 1996.

[09] Y. Nodera, "Global Satellite Mobile for Personal Communications-Integrating Operation with IMT-2000," IIR Global Summit: Third Generation Mobile Communications and IMT-2000, Tokyo, March 1998.

CHAPTER 8

PERSONAL MOBILITY AND UNIVERSAL PERSONAL TELECOMMUNICATION (UPT)

8.1 INTRODUCTION

The cellular mobile and cordless telephony systems described in the preceding chapters provide so-called *terminal mobility*. Terminal mobility systems are characterized by their ability to locate and identify a mobile terminal as it moves and to allow the mobile terminal access—even while it is in motion—to telecommunication services from any location. Terminal mobility is associated with wireless access and requires that the user carry a wireless terminal and be within the radio coverage area. Though the relationship between the network (line termination) and the terminal in mobile networks is a dynamic one, the relationship or association between the terminal and the user is still a *static* relationship. In other words, since the communication is always between the network and the terminal, the call delivery and the billing always must be based on the terminal identity or mobile station number.

Personal mobility, on the other hand, relies on a *dynamic* association between the terminal and the user, with the result that call delivery and billing can be based on a personal identity (personal number) assigned to a user. Personal mobility systems, therefore, are able to identify end users as they move and to allow end users to originate and receive calls, and access subscribed telecommunication services on any terminal, in any location. The emerging implementations and integration of the intelligent network (IN) capabilities within the fixed and mobile networks provide the underpinnings to enable a dynamic relationship between the terminal and the user. With such a dynamic association, complete personal mobility within networks and across multiple networks can be achieved.

8.2   UNIVERSAL PERSONAL TELECOMMUNICATION (UPT): CONCEPT AND SERVICE ASPECTS

8.2.1   The UPT Concept

UPT is a service concept that will deploy the emerging IN capabilities to provide personal mobility to the end users. In the UPT service environment, each user will be assigned a unique personal number (UPT number), which will be dialed by calling subscribers to reach the UPT user. The UPT number also will serve to identify the UPT user at the time of a service request by the UPT subscriber (e.g., call origination or service profile modification). Through manipulation of the service profile, the UPT user will be able not only to designate specific terminals (fixed or mobile) for call delivery and call origination but also to invoke such subscribed supplementary services as call screening and call forwarding. The specific services and features available to UPT users as they roam across different networks and use different terminals will depend on the capabilities of the terminal as well as the capabilities of the network serving the terminal. The key characteristics of UPT include the following:

- terminal and network transparent user identification based on the UPT numbers assigned to individual UPT users
- personal call management, which enables a UPT user to screen incoming calls and control their disposition (based on the identity of the caller, the basic telecommunication service type, the call urgency, time of day, etc.)
- billing of calls on the basis of user's UPT number instead of the identity of the terminal/line deployed by the user for outgoing calls
- subscriber/user control of service profile and personalization of terminal configuration
- service availability/accessibility across multiple network types (e.g., ISDN/PSTN, PSPDN, PLMN) and different terminal types (fixed, portable, mobile)
- network and access security, and confidentiality of user information through UPT user registration and authentication procedures

8.2.2   UPT Service Features

UPT service features represent the set of features required to support the basic operation of UPT, which requires manipulation of UPT service profile data related to personal mobility (i.e., registration data for incoming and outgoing calls and features implicitly needed for this). These UPT service features include the following:

- *Personal numbering*: a UPT number that uniquely identifies each UPT user and is used by a caller to reach that UPT user. A UPT user may have more than one UPT number for different applications (e.g., one for business, another for personal use).

- *Use of UPT service profile*: a personalized service profile that contains data relating to services and options available to the user, which the user can access and manage.

- *Personal charging*: charging associated with the UPT number irrespective of the terminal or network used by the UPT user to originate a call.

- *Single-source billing*: a UPT subscriber will receive bills only from the *home* UPT service provider regardless of whether facilities of different network/service providers have been used.

- *Access security*: authentication/verification of UPT user's identity as required.

- *UPT service profile interrogation*: ability of the UPT users to interrogate their service profiles for purposes of reviewing the status of service and subscription data.

8.2.3   UPT Service Profile Parameters

The UPT service profile is a record containing all the information related to the UPT user that is necessary to provide him or her with the UPT service. UPT user service profiles are associated with UPT numbers assigned to each UPT user. UPT users can access their own service profile data only, and can modify only the part of data mutually agreed between the UPT user and the UPT service provider. The parameters included in the UPT service profile can be divided into two basic categories: fixed parameters and variable parameters from the UPT user's viewpoint. The fixed parameters, typically decided at the service subscription time, include information of the following type:

- UPT number
- default charging reference location (e.g., home location) of the UPT user
- list of services provided by the service provider
- bearer/tele/supplementary services subscribed by the user
- maximum number of failed authentication attempts before access is denied
- type(s) of authentication procedure(s)
- security option(s)

- charging option(s)
- maximum credit limit (if any)
- restrictions on roaming (if any)

Certain parameter types can generally be modified by the UPT user as part of the user service profile management. The information in this category is related either to the service features or to the mobility features. Some typical parameters in this category are as follows:

- restriction on permitted callers (if available)
- authentication procedure currently activated
- activation status of supplementary services
- temporary home location
- default terminal access for incoming calls
- default terminal access for outgoing calls
- variable parameters for routing of incoming calls:

  routing by area/network of call origination

  routing by calling line identity

  routing based on time of day, day of week, and so on

  routing by *on-busy* or *no-answer* condition

8.3   FUNCTIONAL ARCHITECTURE FOR UPT

While the initial implementations of UPT may be provided by using simple database arrangements within the current network capabilities, the evolving IN architecture is better suited for providing many of the signaling and database features required for UPT. Some of these features are:

- user-controllable and portable service profile
- efficient and flexible number translation and call routing and screening
- personalized charging and single-source billing
- rapid service provisioning and alteration
- interworking between service providers

Therefore, as the IN capabilities are increasingly introduced in the networks, many of the functions required to support UPT will be provided within the IN architectural framework. Further, to ensure network and service efficiencies and to implement many of the optional and advanced UPT service features, availability of IN capabilities in the network will be essential. Figure 8.1 illustrates the UPT functional architecture that utilizes the architecture and capabilities of IN CS-1 to support the initial set of UPT services and features, that is, UPT Service Set 1.

Based on the limitations of IN CS-1, the architecture of Figure 8.1 assumes the following for the support of UPT Service Set 1 services and features:

- The interconnection of different UPT-supporting networks is through the SCF–SDF relationship.

- SDFh (SDF home) stores all data related to the UPT user. The UPT data thus are centralized and there is no transfer of the UPT user's service profile information to other databases. This transfer capability will be available with IN CS-2. Thus SDFh needs to be accessed to query or update UPT user's data.

- As a consequence of the preceding constraint, SDFh must provide access control functions to check authenticity of incoming requests, perform UPT user authentication, and maintain all data related to multilateral service agreements, local security measures, and management of the UPT service in the network.

- SCFo (SCF originating) provides the required service logic programs (SLPs).

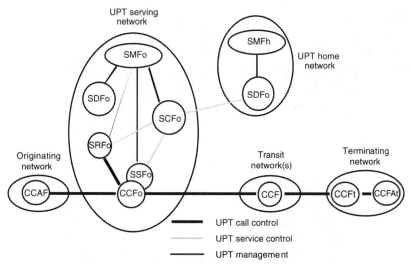

Figure 8.1 UPT functional architecture.

8.4 NUMBERING, ROUTING, AND BILLING ASPECTS

8.4.1 UPT Number (UPTN) and Personal User Identity (PUI)

As mentioned earlier, the basic requirement for UPT service is the assignment to a UPT subscriber of a dialable *personal number*. The UPT number is used to support such key functions as call routing and delivery to a UPT subscriber (incalls), transactions between the UPT subscriber and his or her service profile (e.g., incall and outcall registration), billing of calls made by a UPT subscriber, and identification of UPT subscriber's service profile and his or her home UPT service provider. At this time the UPT service is primarily defined for operation within the context of public switched networks that utilize numbering and addressing based on the ITU-T Recommendation E.164 numbering plan. Examples of these networks include PSTN, ISDN, and PLMNs. A detailed description and discussion of numbering and identities for wireless and personal communications is provided in Appendix A.

The various options for structuring national and international UPT numbers are contained in ITU-T Recommendation E.168, "Application of E.164 Numbering Plan for UPT" [06]. One of the options that is likely to be used in the initial implementations of UPT service is shown in Figure 8.2. In this scheme the international UPT number has the same structure as an

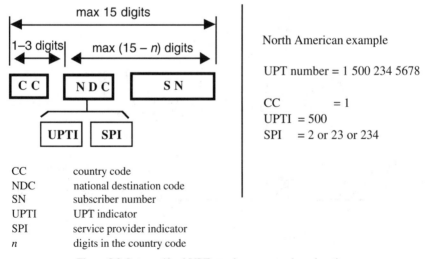

CC	country code
NDC	national destination code
SN	subscriber number
UPTI	UPT indicator
SPI	service provider indicator
n	digits in the country code

Figure 8.2 International UPT number: country-based option.

E.164 number (i.e., CC + NDC + SN). The structuring of the national destination code (NDC) is a national option. However, it is desirable to establish an NDC structure that would allow the calling party and the originating network to recognize that the called number is a UPT number. It is also desirable to include the identity of the service provider (SP) within the NDC to facilitate routing, billing, and accounting of UPT calls. Thus the NDC digits may be partitioned into a UPT indicator field and a service provider indicator field.

Besides the UPT number, a personal user identity (PUI) may be used to facilitate the operation of a UPT service. A PUI unambiguously identifies the UPT user but is different from the UPT number, although there is a one-to-one mapping between them. Thus PUI is an identity that is used internally by the UPT service provider to identify a specific user, and for identifying the user's home UPT service provider to other (visited) UPT service providers and networks supporting UPT. PUI may also be used, in association with a PIN (personal identification number) for UPT user–network authentication. PUI used in UPT is analogous to the international mobile subscriber identity (IMSI) that is used in many cellular networks. In fact, the structure of PUI is also based on Recommendation E.212, which defines the structure for IMSI.

8.4.2   Outcall Registration and Incall Routing and Delivery

As a UPT user roams across different networks and attempts to originate calls, it is necessary to register (current location) with the home UPT service provider or network. This registration or *outcall* procedure basically consists in the user updating service profile (SP) information that resides in a database in the home network. The initial part of the registration process may also involve some type of *user authentication* to ensure the legitimacy of the registering UPT user in terms of the user being a legitimate user of the service. Once the user has registered, he or she can originate calls, which will then be billed against the person's UPT personal number.

For calls directed to the UPT user, the originating network will first interrogate the user's service profile (resident in the called UPT user's home network database), which will return a *routing address* based on the current location (registration) of the called UPT user. The call is then routed and delivered to the called UPT user based on the routing address. Optionally, at the request of the calling user, the terminating network may invoke an authentication procedure to ensure that it is the actual called UPT user who is accepting the call (person-to-person calling feature). An example of outcall authentication/ registration and incall routing/delivery is illustrated in Figure 8.3, where it is

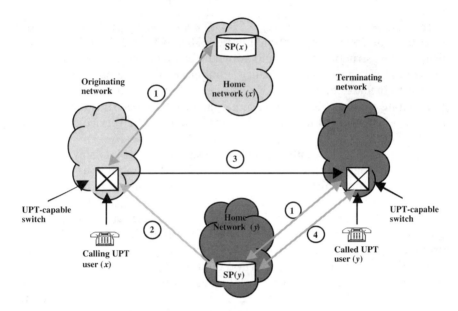

1 Authentication/Registration of UPT users *x* and *y*.
2 Retrieval of access and service information for UPT user *y*.
3 Call routing from terminal *x* to terminal *y*.
4 Authentication of UPT user *y* (optional).

Figure 8.3 Example of UPT outcall registration and incall routing.

assumed that both the originating and terminating networks are UPT capable (i.e., they have technical capabilities to access and interrogate remote databases). Of course, the visited and the home networks need to have bilateral roaming agreements.

8.4.3 Charging and Billing Aspects of UPT

Charging and billing for UPT services is associated with the unique UPT number assigned to a subscriber. Even though UPT users access services from different terminals served by different networks in different countries, it is expected that each subscriber will receive a single bill from his or her home service provider. The charges incurred by a UPT user fall into the following categories:

- subscription-related charges (e.g., monthly charges for basic and supplementary services)

- subscription-management-related charges (e.g., for service profile modifications)

- call-related charges, which depend not only on duration and distance but also on the location of the called UPT user relative to his or her home location.

Whereas in the PSTN the calling party is entirely responsible for call charges, in mobile networks the charges may be split between the calling and called parties depending on the location of the called party at the time of a call. Generally the calling party is charged for the part of the call path to called party's home location, while the called party is charged for the additional call path in a roaming situation. Similarly in the case of UPT, location-related split charging is applicable. In other words, the actual charges to the calling and called parties depend on the location of the two parties. The possible scenarios depend on the location of the visited network relative to the location of the calling party, and these are summarized in Table 8.1.

Table 8.1 Possible Scenarios for Split Charging in UPT

Called User Registration	Total Charge	Charge to Calling User	Charge to Called User
Home network	Cn (normal)	Cn	None
Visited network-1	Cn + C1	Cn	C1
Visited network-2	Cn − C2	Cn − C2	None

8.5   SCENARIOS FOR PARTITIONING AND LOCATION OF SERVICE PROFILE INFORMATION

It is generally assumed that the entire set of service-profile-related information for UPT users will reside in their *home networks*. This assumption will generally be valid in the early phase of UPT implementation when the UPT databases will be very few and more centralized (e.g., one per service/network provider). However, from the perspective of routing efficiency, quality of service, and signaling network performance, it may be desirable to use more distributed architectures, where the service profile information/data can be partitioned and/or duplicated and distributed over different locations. For this purpose the service profile may be divided into categories, as shown in Figure 8.4.

Figures 8.5, 8.6, and 8.7 illustrate the possible scenarios for distributing the service profile information between the home and visited networks with asso-

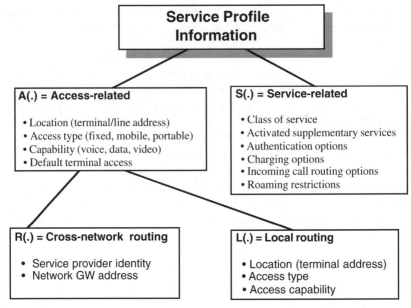

Figure 8.4 Subdivision of UPT service profile information.

ciated impact on signaling traffic loads. In the scenario of Figure 8.5, the entire service profile for a UPT user resides at the home network database. This scenario will require database queries across the largest segments of internetwork signaling links, thus larger delay in the signaling network and call setup. The following notations apply in Figures 8.5–8.7:

$$Ax/y = \text{access-related information for UPT user } x/y$$

$$Sx/y = \text{service-related information for UPT user } x/y$$

where 1 = authentication/registration of UPT user x or y
2 = retrieval of access and service information for UPT user y
3 = call routing from terminal x to terminal y
4 = authentication of UPT user y (optional)

In the second implementation (Figure 8.6), the service profile is partitioned into access-related (Ax/y) and service-related (Sx/y), when the UPT user indicates his or her intention to visit another network (i.e., invokes *registration* procedures from a visited network), the service-related information is copied into the visited network database. As seen in Figure 8.6, such distribution of information can significantly reduce cross-network signaling traffic flows. Such an architecture is especially desirable for medium- and long-term cross-network visitors.

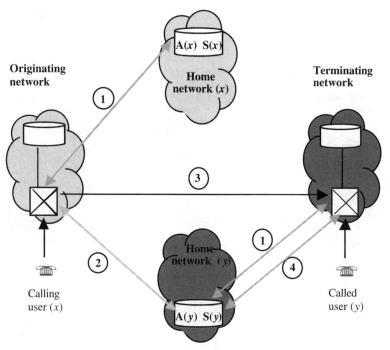

Figure 8.5 Complete service profile at the home network database.

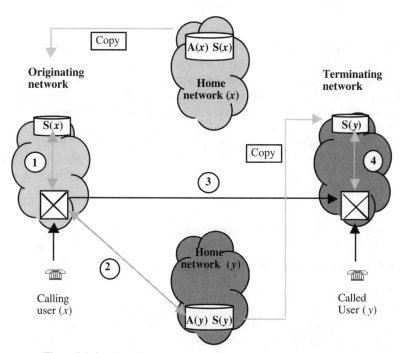

Figure 8.6 Service-related information at the visited network database.

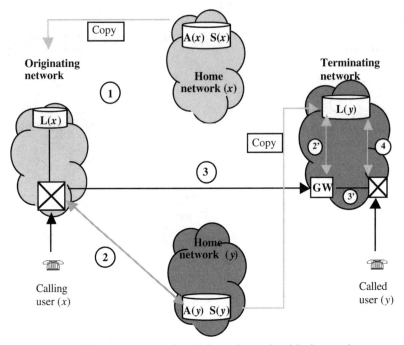

Figure 8.7 Partial access-related information at the visited network.

Further reductions in cross-network signaling traffic (and thus call setup delay) can be achieved by further partitioning the access-related information Ax/y into the "basic" network routing information R(.), which will identify the service provider (SP) and the nearest gateway (GW) in the terminating network to which the call needs to be routed, and the remaining detailed local access-related information L(.), which is required to route the call within the terminating network, the latter being copied to the visited network only. This scenario is illustrated in Figure 8.7, where 2′ = retrieval of detailed access-related information for call completion and 3′ = call routing in the visited network.

8.6   ACCESS SECURITY REQUIREMENTS FOR UPT

8.6.1   Access Security: General Aspects

UPT access security through some form of UPT user authentication/ verification procedure is essential to protect the user's resources or information from unauthorized persons (fraudulent use) and/or from unauthorized listening or recording of information (eavesdropping). The access security procedures thus minimize the probability of incorrect charging and malicious call redirection, and ensure the integrity of users' information and privacy. The UPT service/network provider also needs access security procedures to maintain integrity of billing and operational information. There are essentially three logical operations for secure UPT access:

1 *UPT user recognition*, which ensures that the network will recognize and separate UPT users (on service request) for further authentication and verification.

2 *UPT user authentication*, which ensures that the access to the UPT service is restricted to legitimate (authentic) users only.

3 *UPT user validation*, which ensures that the personal identity claimed by the user really belongs to the user.

However, in a specific implementation, one or more of these logical operations may be combined into a single operation. The choice of a secure UPT access method will be dictated by a number of factors, which include the following:

- level of protection required (weak or strong authentication)
- applicable range of terminal types (rotary, DTMF, ISDN, mobile)
- simplicity and user-friendliness of the method
- level of standardization required
- cost to user/service provider

8.6.2   UPT Access Security: Implementation Options

8.6.2.1   Schemes Based on a Personal Identification Number.   The use of a (personal identification number (PIN) and DTMF signaling is similar to the use of ID and password familiar from computer or data network access. The user identity in the case of UPT user authentication can be based either on a UPT access number (UPTAN) uniquely assigned to each UPT user by the home service provider (UPTAN is in addition to and different from the UPT personal number) or on a UPT access code (UPTAC) assigned to each UPT user. The role of UPTAN and UPTAC in supporting UPT procedures is described in ITU-T Recommendation E.168 [06]. The level of security provided by this method is comparable to the public data service access arrangements. Frequent password/PIN changes are recommended/enforced, and the resulting level of access security is generally considered adequate for public communication services.

8.6.2.2   Schemes Based on a UPT Device (Smart Card).   A higher level of protection and access from a wider variety of terminals may be achieved by using a tone-type UPT device. The device requires the UPT user to be authenticated to the device (e.g., by entering a password/PIN into the device to arm it before each use) to prevent unauthorized parties from fraudulently using the device.

A so-called smart card will use a standardized audio signaling protocol to communicate with the local UPT service provider through a voice channel. It must perform necessary calculations for secure authentication of the UPT user's identity and for authentication of the requested procedure. The algorithms can be standardized to permit remote authentication across networks.

The UPT user will interact with the device to enter the desired procedures, together with any associated data. Upon completion of this interaction, the UPT user will dial a UPT access code to connect with the local UPT serving node. The UPT user would then use the UPT device to send a burst of tones. The UPT serving node will respond with appropriate confirmation messages or an indication of reasons for denial of service.

8.6.2.3   Schemes Based on Speech Recognition.   The requirements of secure UPT access can be met in a simple, user-friendly, and ubiquitous manner by the application of the rapidly maturing speech recognition technologies. To provide highly secure UPT service access, speech recognition technology requires the following capabilities.

1 *Speech Recognition*: the ability for the network to automatically identify a set of user commands and parameters (e.g., directory numbers, passwords, user names) input as spoken words over a typical telephone terminal/line (UPT user recognition). This capability is not necessarily deployed in all cases, instead a dialed UPT prefix, a UPT access code, or the UPT number may be used to gain access to the UPT server.

2 *Speaker identification*: labeling an unknown voice/speaker as one of a set of known (recorded) voices/speakers. This capability is used to identify whether a UPT user requesting service is one of the authorized users (UPT user authentication).

3 *Speaker verification*: determining whether an unknown voice matches the known voice of a speaker whose identity is being claimed, and is used to verify the personal identity of a specific UPT user (UPT user verification).

The location and handling of the *speech template* for speaker identification/verification for UPT service access is an important network architecture/design consideration. As indicated above and illustrated in Figure 8.8, in the first phase, the UPT user will gain access to the local UPT service provider

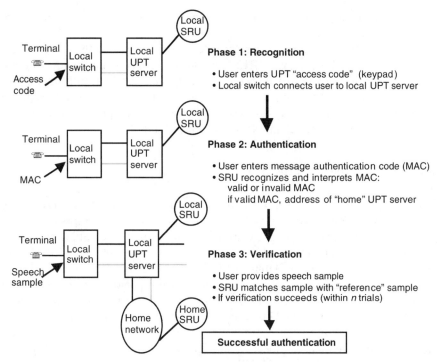

Figure 8.8 UPT access security using the speech recognition method.

through the serving local exchange—the UPT server being capable of accepting both voice input and input through tone dialing.

In the second step, the UPT user inputs an identification code, which can be a set of numbers (or a name/phrase) spoken into a speech input device. The speech recognition unit (SRU) in the local UPT server interprets the speech input, and the user authentication procedure proceeds in a manner similar to credit card verification. Where necessary, the local UPT server can also identify the calling user's home UPT server (from the authentication information provided), where the *speech template* of the calling user may be stored for speaker verification.

In the third step, the UPT user provides a speech sample consisting of either a predetermined phrase, or one of the predetermined phrases on appropriate prompting from the local SRU. The latter provides a better level of security against breach through eavesdropping/recording, though it does require more storage and complex SRU operation.

Various options are possible and need to be assessed for the final or speaker verification phase. The type, location, and handling of the speech template will have an impact on level of security, routing and network efficiency, signaling traffic loads, and overall system performance and quality of service delivered to the user.

It is expected that the user's reference speech sample or template will reside at the home UPT server/SRU. Thus the key aspects that need to be addressed include the site at which the actual speech verification (matching of speech sample with the template or reference sample) takes place [i.e., at the home SRU or local (visited) SRU] and how the two samples are brought together (i.e., using bearer channels or signaling channels).

8.7 CONCLUDING REMARKS

Personal mobility services offer the user the option of originating and receiving calls on any terminal (fixed or mobile) at any time. Earlier attempts to provide some degree of personal mobility are represented by use of calling cards (for call origination and billing), and by paging and personal number services (for incoming calls). Using the availability of ISDN and IN capabilities in today's networks, UPT consolidates all the requirements for personal mobility service within a single set of service and network specifications. These specifications have been developed by the ITU-T as well as by ETSI. In North America, many of the personal mobility concepts have been included in PCS, which, by definition, is assumed to include aspects of both personal and terminal mobility.

In terms of availability of UPT or personal numbering services in various parts of the world, the implementation of these services by network and service providers has been slow and sporadic. In the United States AT&T provides these services under the *True Connections* label and allows calls to be diverted to most numbers in the country. The *personal numbers* are allocated in the 500 Service Access Code assigned for such services (Appendix A). UPT services are also available in Europe (e.g., Norwegian Telecom) and in Japan (NTT). However, the penetration of these services is relatively low compared with terminal mobility services (mobile and personal communication services).

However, the assignment and use of a *personal number* is a very attractive feature. Not only does it ensure that a call can be delivered to the user at any location, but it also provides complete number portability (across service providers as well as across geographic locations). It is expected that the market for UPT and *personal number* services will grow as a complementary service to *terminal mobility* services, rather than as a stand-alone service.

8.8 REFERENCES

[01] M. Zaid, "Personal Mobility in PCS," *Personal Communications Magazine,* Vol. 1, No. 4, 1994.

[02] R. Pandya, "Network Architecture and Functional Requirements for UPT," INFOCOM 92, Florence, Italy, May 1992.

[03] ITU-T Recommendation F.850, "Principles of Universal Personal Telecommunication (UPT)," International Telecommunication Union, Geneva, 1992.

[04] ITU-T Recommendation F.851, "Universal Personal Telecommunication (UPT) Service Description—Service Set 1," International Telecommunication Union, Geneva, February 1995.

[05] ITU-T Recommendation F.852, "Universal Personal Telecommunication (UPT) Service Description—Service Set 2," International Telecommunication Union, Geneva, 1997.

[06] ITU-T Recommendation E.168 (Revision 1), "Application of E.164 Numbering Plan for UPT," International Telecommunication Union, Geneva, May 1999.

[07] ITU-T Recommendation Q. 1541, "Procedures for Universal Personal Telecommunication Functional Modeling and Information Flows," International Telecommunication Union, Geneva, 1999.

[08] R. Pandya, "Traffic Performance and QOS considerations for UPT," Eighth ITC Specialist Seminar, Genoa, Italy, October 12–14, 1992.

[09] M. Fujioka, S. Sakai, and H. Yagi, "Hierarchical and Distributed Information Handling for UPT," *Network Magazine*, Vol. 4, No. 6, November 1990.

[10] R. Pandya, "Universal Personal Telecommunications (UPT)—Service Features, Network Aspects and Performance Considerations," TRC Report No. 3/1992, Teletraffic Research Centre, University of Adelaide, Adelaide, Australia, February 1992.

APPENDIX A

NUMBERS AND INDENTITIES FOR MOBILE AND PERSONAL COMMUNICATION SERVICES

A.1 INTRODUCTION

Suitable *numbers* and *identities* represent key requirements for mobile and personal communication services because they are needed to support such basic functions as addressing, call/connection setup, location management, registration, call delivery, and charging and billing. In a fixed network, functions related to mobility management are not required, and the call/connection setup functions can be achieved through the use of subscriber/terminal (directory) numbers. However, in mobile and personal communication networks, well-defined and standardized user/terminal identities are required to efficiently handle the mobility management functions. Some form of *mobile station equipment identities* are also needed to prevent the use of non-type-approved or stolen terminals in the network.

ITU-T has developed international standards on mobile network numbering and identities. With increasing demand for national and international roaming capability, these standards for numbering and identities either have been adopted and implemented (e.g., in GSM) or have been adopted and are awaiting implementation (e.g., in IS-136). Further, the ITU-T is enhancing the scope of these standards to accommodate new mobility services like UPT, GMPCS, and IMT-2000. Parallel activity in North American forums is addressing the application, management, and administration of numbering and identity resources for emerging wireless and PCS networks. Figures A.1 and A.2 identify the key international (ITU-T) and North American forums, respectively, which are involved in various aspects of numbering and identities for mobile and PCS networks. The major European activity on numbering related standards takes place in ETSI's Technical Committee (TC) SPAN2.

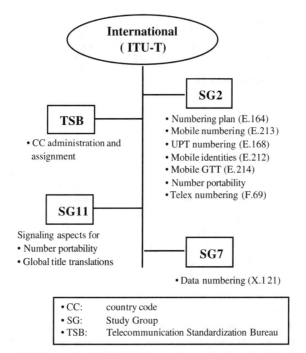

Figure A.1 International (ITU-T) forums for numbering-related activities.

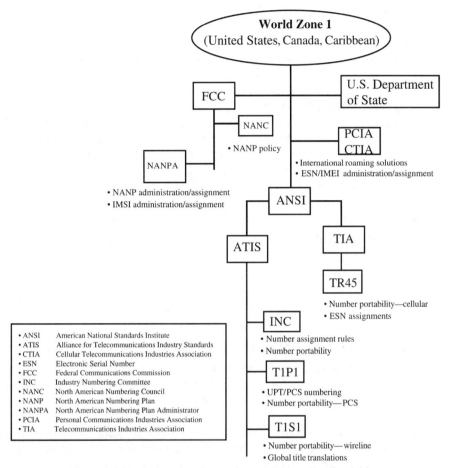

Figure A.2 North American forums for numbering-related activities.

A.2   ROLE OF MOBILE/PCS STATION NUMBERS AND IDENTITIES

Directory numbers (DNs) are assigned to fixed and mobile stations as dial-able, public numbers conforming to the E.164 international numbering plan, and they are used to support the following functions:

- call routing to the destination terminal or to the MSC/HLR of the called mobile user
- charging and billing management

- dialing/addressing for call origination
- calling/called party identification
- providing location and/or service information to the users

E.164 numbers are also used in mobile networks as *roaming numbers* provided by the visiting network to the home network for call setup (call routing) to a roaming subscriber. These roaming numbers are used by the network (not dialed by the user) and are not assigned to mobile stations as directory numbers.

Mobile station identities are assigned to mobile stations and are used internally by the mobile network to support such mobility management functions as the following:

- determination of the home network or HLR of a roaming terminal
- identification of a mobile station (MS) on the radio control path for location update/registration
- identification of the MS for all signaling on the radio control path (e.g., for paging)
- identification of the MS when information about the MS is to be exchanged within the mobile network(s)
- identification of the MS for charging and billing of roaming terminals
- identification of the MS for updating or retrieving subscriber profiles that may contain such items as subscribed (switch-based) features and triggers to query service logic for enhanced services

To support international routing of calls and international roaming of mobile subscribers, the mobile station numbers and identities must conform to internationally standardized formats.

Terminal equipment identities are also used in mobile networks to ensure that the mobile terminals being served by the public mobile network are legitimate (type-approved) and are not stolen or fraudulent terminals. Rather than being based on an internationally standardized format, however, these equipment identities are defined, managed, and assigned on a national or regional basis.

A.3   INTERNATIONAL STANDARDS ON NUMBERING AND IDENTITIES

International (ITU-T) recommendations on numbering and identities that are relevant for mobile and personal communication services are summarized in Table A.1.

Table A.1 ITU-T Recommendations on Numbering and Identities

ITU-T Recommendation	Recommendation Title	Scope
E.164 (revised)	International Public Telecommunication Numbering Plan	Basic numbering plan for public switched networks (PSTN/ISDN/PLMN)
E.212 (revised)	Indentification Plan for Mobile Terminals and Mobile Users	Mobile station/subsccriber identities (IMSI)in PLMNs and personal user identities (PUI) for UPT
E.213	Telephone and ISDN Numbering Plan forLand and Mobile Stations in PLMN	Scenarios for application of E.164 numbers for PLMNs
E.214	Structure of the Land Mobile Global Titlefor the SCCP	Provides mapping of E.212 identities of roaming subscribers for SCCP routing
E.168 (revised)	Application of E.164 Numbering Plan for UPT	Scenarios for using E.164 numbers for personal mobility (UPT)

A.3.1   The International Public Telecommunication Numbering Plan: ITU-T Recommendation E.164

Recommendation E.164 is the primary standard that specifies the international numbering structure for circuit-switched networks and services. The recommendation addresses numbering structures for geographic areas (e.g., countries and zones), for global services (e.g., the International Freephone Service), and for global networks (e.g., GMPCS). The format of an E.164 (international) number for geographic areas (Figure A.3) consists of three parts:

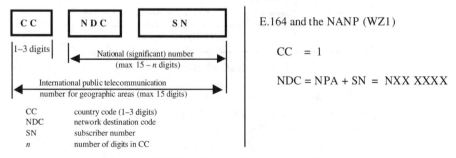

Figure A.3 E.164 numbering plan for geographic areas.

the country code (CC), the network destination code (NDC), and the subscriber number (SN). The North American Numbering Plan (NANP), which is used in World Zone 1 (WZ1) conforms to E.164 as shown in Figure A.3. In WZ1 the NDC may consist of the NPA plus additional digits from the NXX (central office code).

To determine the destination country, the most appropriate network routing, and proper charging, the originating country must analyze a number of digits of the international (E.164) number. For international outgoing calls, the number analysis performed at the originating country need not be more than the country code of the destination country (or zone) and

> four digits of the national significant number in the case of a country with a three-digit CC

> five digits of the national significant number in the case of a country with a two-digit CC

> six digits of the national significant number in the case of a country with a one-digit CC

A.3.2   Identification Plan for Terminals and Mobile Users: ITU-T Recommendation E.212

This recommendation defines an international identification plan for land mobile stations designed to enable mobile stations to roam among public land mobile networks located in different countries. The structure of international mobile station identities (IMSIs) based on E.212 is shown in Figure A.4.

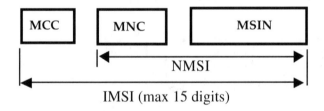

MCC   mobile country code
MNC   mobile network code
MSIN   mobile station identification number
NMSI   national mobile station identity
IMSI   international mobile station/subscriber identity

Figure A.4 IMSI structure according to Recommendation E.212.

For assignment of IMSIs, Recommendation E.212 provides the following principles:

- Only numerical characters (0–9) should be used.
- IMSI should not exceed 15 digits.
- MCC consists of 3 digits and is assigned by ITU-T.
- NMSI is assigned and administered nationally.
- Allocation of MNCs should ensure that no more than 6 digits of the IMSI need to be analyzed in a visited network for information transfer (e.g., to the HLR).
- It is desirable that the mobile station identities plan (E.212) not be directly related to the mobile station numbering plan (E.164/E.213) in use in the country/zone.

Note 1   The ETSI-specified GSM/DCS1800 system and its derivative PCS1900 in North America use IMSI based on E.212. However, these systems use the term "international *mobile subscriber* [rather than *mobile station*] identity."

Note 2   To provide confidentiality and reduce signaling load over the air interface in these systems, the VLR may allocate a unique temporary mobile station identity (TMSI) to the visiting MS while maintaining a correlation with the IMSI for the MS.

Note 3   The original Recommendation E.212 has been revised and provides the basic structure (Figure A.4) for personal user identity (PUI) for use in universal personal telecommunication (UPT) and subscriber identities for global mobile personal communications via satellite (GMPCS).

A.3.3   Numbering Plan for Mobile Networks: ITU-T Recommendation E.213

Recommendation E.213 provides the principles for application of E.164 numbers for land mobile services. It addresses the assignment of numbers to mobile stations as well as the assignment of *roaming numbers* for call delivery to stations roaming outside their *home* service areas. Recommendation E.213 specifies the following national options for assignment of numbers to mobile stations:

1. Use of integrated numbering. In this option, mobile station numbers are fully integrated with the national telephone/ISDN numbering plan. For example, all cellular networks in North America utilize this option, and in this option it is possible that the calling user and the originating network will not recognize that the called party number is a mobile station number.

2. Use of special service access or trunk code. In this option, the public mobile network/service may be regarded as a separate numbering area (NPA/trunk code) within the national telephone/ISDN numbering plan. This option is deployed in many cellular mobile networks outside North America, where a special trunk code is assigned for cellular mobile service. In this option, the calling user and the originating exchange can recognize that the called party number is assigned to a cellular mobile station.

Recommendation E.213 also addresses the allocation of *roaming numbers* to route calls to mobile stations that are roaming outside their home networks. These roaming numbers form part of a pool of reserved numbers in the numbering plan of the *visited network*. These can be allocated at the time the mobile station roams into the visited network (and stored in the visited subscriber's

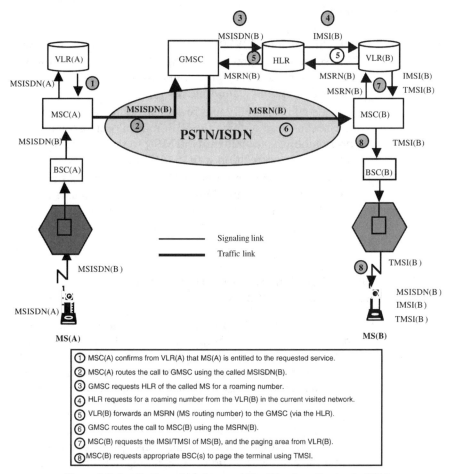

① MSC(A) confirms from VLR(A) that MS(A) is entitled to the requested service.

② MSC(A) routes the call to GMSC using the called MSISDN(B).

③ GMSC requests HLR of the called MS for a roaming number.

④ HLR requests for a roaming number from the VLR(B) in the current visited network.

⑤ VLR(B) forwards an MSRN (MS routing number) to the GMSC (via the HLR).

⑥ GMSC routes the call to MSC(B) using the MSRN(B).

⑦ MSC(B) requests the IMSI/TMSI of MS(B), and the paging area from VLR(B).

⑧ MSC(B) requests appropriate BSC(s) to page the terminal using TMSI.

Figure A.5 Application of numbers and identities for call setup in GSM.

HLR) for as long as the mobile station remains in that visited network; alternatively, the roaming number may be allocated by the VLR on a call-by-call basis (on request from the HLR). The latter method is more common: both AMPS-based and GSM cellular networks utilize this approach. Figure A.5 illustrates the application of mobile station numbers (MSIN), IMSI, and roaming numbers (MSRN) for call delivery to a roaming mobile station in a GSM network.

A.3.4   Structure for the Land Mobile Global Title: ITU-T Recommendation E.214

To permit roaming between public land mobile networks (especially mobile networks located in different countries), it is necessary to transfer information (over the SS7 signaling network) between various mobile network elements—for example, transfer of a roaming number or the mobile station identity (IMSI) between a VLR and an HLR.

When a mobile station roams into a foreign country (or region with a different E.164 CC mobile network), it needs to register with a VLR in that network. The only information available to the VLR to address the mobile station's HLR is its IMSI. The SS7 network, however, can currently route messages only on the basis of E.164 numbers or point codes and subsystem numbers (PC/SSN)—the latter do not generally work across international boundaries.

Recommendation E.214 defines the relationship between a mobile global title (MGT) and the IMSI. The MGT so derived from the IMSI has an E.164-like structure and can be used for addressing/routing messages (over the SS7 signaling network) to the mobile station's HLR for registration. The relationship between the IMSI and the MGT is shown in Figure A.6.

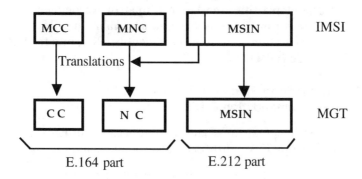

- The MGT is of variable length with a maximum of 15 decimal digits.
- The SCCP analyzes only the E.164 part for addressing and routing.
- The CC in the MGT permits the identification of the home country of the roaming MS.
- The (NC) in the MGT identifies the mobile network or the HLR of the roaming MS.
- The E.212 part is used to identify the MS or MS and HLR.

Figure A.6 Relationship between IMSI and MGT.

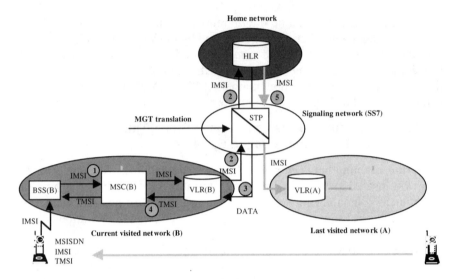

① MS provides its IMSI to the VLR(B) of the current visited network.
② The VLR(B) sends the IMSI of the roaming MS to the HLR; STP/SCCP routes the message using MGT.
③ The HLR returns the subscription data and the authentication and encryption triplets to the VLR.
④ The VLR assigns a TMSI to the MS.
⑤ The HLR informs the VLR(A) of the last visited network that the MS has left its jurisdiction using IMSI.

Figure A.7 Application of IMSI and TMSI for location update in GSM.

Figure A.7 illustrates the application of the MGT based on Recommendation E.214 for location updating or registration of a roaming mobile station across GSM networks.

Note 1   The global title translation format of Recommendation E.214 shown in Figure A.7 is also applicable in UPT networks for translation of personal user identity (PUI) into the UPT global title (UPTGT).

A.3.5 Numbering Plan for UPT: ITU-T Recommendation E.168

The following basic options or schemes for application of E.164 numbers for personal mobility (UPT) are provided in Recommendation E. 168:

home-network-based scheme, where the number does not contain any special UPT identity recognizable from outside the specific national (UPT-supporting) network

country-based scheme, where the NDC is used to identify a UPT call as well as the (UPT) service provider

global scheme, where a country code (878) is assigned to UPT service

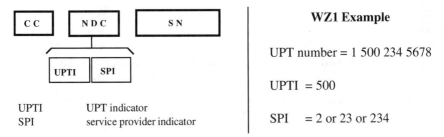

Figure A.8 UPT numbering: country-based option.

The numbering structure for the *country*-based scheme that is likely to be used in initial UPT implementations is shown in Figure A.8. The international UPT number has the same structure as an E.164 number (i.e., CC + NDC + SN), though a national UPT number in this case must always include the NDC. Though the NDC structuring is a national option, it is desirable to establish a structure that would allow the calling party and the originating network to recognize that the called number is a UPT number. It is also desirable to include the identity of the service provider (SP) within the NDC to facilitate routing, billing, and accounting of UPT calls.

A.4   USAGE OF MOBILE/PCS NUMBERS AND IDENTITIES IN NORTH AMERICA (WZ1)

A.4.1   Mobile and PCS Station Numbers

As mentioned earlier (Section A.3.3) the mobile numbering plan can be fully integrated with the national PSTN/ISDN numbering plan. The current numbering plan for the cellular networks in North America is fully integrated with the North American Numbering Plan (NANP), and cellular service providers are allocated full or partial central office (CO) codes in the various NPAs in the NANP. Thus the numbering and addressing plan for mobile/PCS stations in WZ1 fully conforms to ITU-T Recommendation E.213 for mobile station numbers. Temporary local directory numbers (TLDNs) are assigned by the visited network (MSC/VLR) on a call-by-call basis to make possible the routing of calls to mobile stations roaming outside their home networks.

Most of the PCS networks currently being implemented are being assigned geographic (NPA NXX) numbers as PCS station numbers in a manner analogous to the cellular networks in North America. However, the 500 Service Access Code (SAC) has been identified for personal communication services in North America, and therefore PCS station numbers can take the form 500

NXX XXXX, where the first three digits (500) will identify the service and the next 3 or more digits will identify the service provider. The deployment of the 500 code for PCS requires implementation of appropriate access arrangements (opening of 500 code) in WZ1. These access arrangements for 500 numbers have been implemented, and the numbers within the 500 code are being used.

A.4.2   Mobile and PCS Station/Subscriber Identities

Currently 800 MHz AMPS-based cellular networks in North America use mobile identification numbers (MINs) as the mobile station identities. The MIN for a mobile station is derived from the 10-digit mobile station directory number (NPA NXX XXXX) as a 34-bit binary sequence by means of a coding algorithm specified in the AMPS standard, as shown in Figure A.9. The MIN structure does not conform to the international E.212 standard, and further it does not follow the E.212 guideline that the MS identity should not be related to its directory number.

Whereas the use of MIN (based on the DN) in North America has certain advantages—separate MS identities do not need to be stored (in the HLR), nor do they need to be allocated and managed—there are some drawbacks to the use of MIN in the emerging mobile communication environment. These include the following:

- difficulty in supporting roaming outside North America (international roaming)
- possibility of conflict with MS identities in neighboring countries (e.g., Mexico)

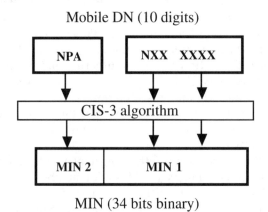

Mobile DN (10 digits)

MIN (34 bits binary)

Figure A.9 Format of MIN currently used in AMPS-based cellular networks.

- automatic change and reprogramming of MIN (into the terminal) induced by a change in the mobile station number (DN)

As discussed in the following sections, efforts are under way to gradually replace the current MIN-based identities by E.212-based international identities.

E.212-based IMSI has been adopted by the TIA for emerging cellular standards like IS-136 (TDMA) and IS-95 (CDMA) as a long-term solution to support international roaming. However, to ensure backward compatibility with existing terminals and base stations with MIN capability only, intermediate solutions will be required. For example, the IS-136 specification allows for a mobile to have both an IMSI and a MIN with the mobile transmitting MIN in the home country, and IMSI outside. Use of TMSI on the digital control channel (DCCH) is also supported in IS-136.

Since the IMSI will also be used in the emerging PCS services (in the 1800 MHz band), there must be sufficient IMSI capability to serve all wireless-based networks/service providers (i.e., cellular, PCS, satellite, and SMR). To this end, it is being proposed that a three-digit mobile network code (MNC) be adopted so that, combined with seven mobile country codes (MCCs) allocated to the United States (by the ITU-T), up to 7000 wireless-based network/service providers can be supported in the United States. The following multifaceted issue, arising out of the adoption of IMSI in North America, is being addressed in the ITU-T:

- WZ1 has a common E.164 country code (= 1), and individual countries within WZ1 are assigned their own E.212 mobile country codes, with seven MCCs (310–316) assigned to the United States. Because of the assignment of multiple mobile country codes within a region with a single PSTN/ISDN country code, the E.214 mapping of IMSIs to mobile global titles (MGTs) shown in Figure A.6 does not produce an unambiguous E.164 address for the roamer's home network or HLR. This ambiguity in the E.214-based mapping between the roaming terminal's IMSI and the MGT is illustrated in Figure A.10.

- In the example shown in Figure A.10, the two mobile networks N1 and N2 in the United States are distinguished from each other by assigning two different MCCs (310 and 311) but the same MNC (123). For countries that have a single MCC, two networks cannot be assigned the same MNC.

- For routing messages over the signaling network, the HLRs in the two networks have E.164 addresses (e.g., HLR1 = 1 415 NXX XXXX, HLR2 = 1 202 NXX XXXX).

- When a mobile station from N1 (with IMSI = 310 123 XXXX) roams into a country outside the United States, the VLR is required to send a

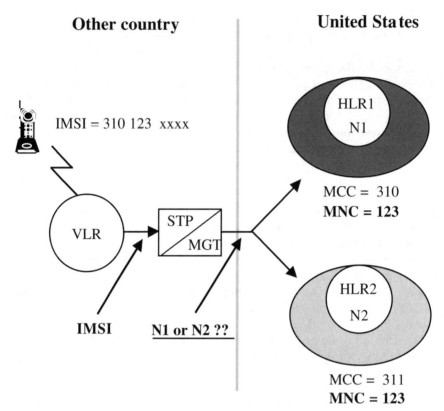

Figure A.10 Impact of multiple E.212 mobile country codes assigned to a region having a single E.164 country code.

message to N1 (HLR1) for location updating, based on the information contained in the visited mobile station's IMSI.

- For routing the message to N1 (HLR1), the IMSI for the visited mobile station needs to be mapped to the E.164 address of HLR1 via E.214 global title translation. However, the mapping of the MCC + MNC (310 123) does not provide a unique E.164 address for the HLR because the mapping can lead to either HLR1 (1 415 NXX XXXX) or HLR2 (1 202 NXX XXXX), since networks N1 and N2 have the same MNC (123).

The effect of having a single E.164 country code in WZ1 is that both MCCs (i.e., 310 and 311) map to the same value of 1 for the CC of the MGT. Thus, for IMSIs to work effectively in WZ1 to support international roaming, it is necessary either to modify E.214 so that the MGT can uniquely identify the home network when multiple MCCs are assigned to a country or numbering zone, or to standardize an additional global title translation, which will route signaling messages to an E.212 format address. This new GTT has now been standardized by Committee T1S1 in the United States, and Study Group 11 of the ITU-T is defining an additional global title translation type, based on the E.212 format, as an international standard.

A.5   TERMINAL EQUIPMENT IDENTITIES

As mentioned in Section A.2, mobile/PCS station equipment identities are used to ensure that only legitimate terminals are registered and allowed service by the network. There are no international standards that specify structures and assignment guidelines for such identities. However, regional or system-specific procedures have been developed. For example, for GSM/PCS1900 this identity is provided by the so-called international mobile equipment identity (IMEI), which has the structure shown in Figure A.11.

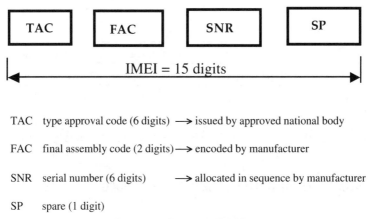

TAC   type approval code (6 digits) —→ issued by approved national body

FAC   final assembly code (2 digits)—→ encoded by manufacturer

SNR   serial number (6 digits)   —→ allocated in sequence by manufacturer

SP   spare (1 digit)

Figure A.11 Structure of IMEI.

In North America, for the AMPS-based cellular (800 MHz) and AMPS-based PCS systems, the mobile station equipment identity is provided by the electronic serial number (ESN). The ESN, a 32-bit binary number is factory-set and not readily alterable in the field. The current bit allocations (structure) of the ESN are shown in Figure A.12.

A manufacturer's (MFR) code is assigned by the assigning authority at the time of initial type approval, and the serial numbers are assigned uniquely by the manufacturer. Till recently the FCC was the assigning au-

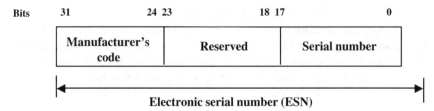

Figure A.12 Composition of an electronic serial number (ESN).

thority for MFR codes in the United States, but the responsibility now lies with the TIA (TR45). The ad hoc group on ESN in the TIA has also been addressing the issue of ESN expansion so that the number of ESNs can be increased to support the forecasted future demands. The proposals are indicated in Table A.2, where the use of the long MFR codes will provide additional ESNs without affecting the 32-bit codes. The expanded ESNs, which represent the long-term solution, will have very significant impact on all North American wireless systems in terms of data entry, signaling, databases, billing, and so on.

Table A.2 Current and Planned ESN Structures

		MFR Code		Serial Number	
ESN Format	**Total Length**	**Bit Range**	**Number**	**Bit Range**	**SNs/MFR**
Current format	32 bits	24–31	251	0–23	16,777,216
Long MFRs	32 bits	18–24	128	0–17	262,144
Expanded ESNs	56 bits	32–47	65,543	0–23	16,777,216

A.6   NUMBERS AND IDENTITIES FOR IMT-2000

As part of the activity of IMT-2000 with respect to signaling requirements and network capabilities, ITU-T is also addressing the nature of numbers and identities that may be required to support IMT-2000 services and features. The mobility aspects associated with IMT-2000 and related numbers and identity needs are illustrated in Figure A.13.

Besides supporting terminal mobility, IMT-2000 will support UPT. The support of UPT by IMT-2000 implies that all UPT features (e.g., incall registration, outcall registration, service profile management, and multiple UPT user registrations on a IMT-2000 terminals for incoming calls) will be supported and will be available to every UPT subscriber. IMT-2000 user mobility, on the other hand, is a form of personal mobility that is restricted to subscribers of IMT-2000 service and is enabled through the use of a easily de-

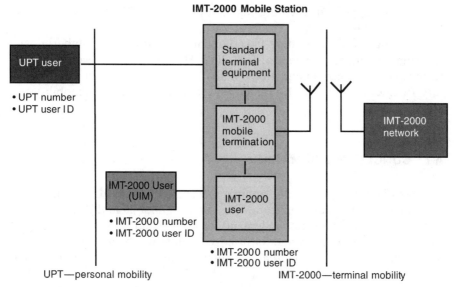

Figure A.13 Mobility aspects of IMT-2000 and related numbers and identities.

tachable and portable device, similar to the subscriber identification module
cards used in GSM.

Table A.3 summarizes current ITU views on types of numbers and identities needed to support the various types of mobility indicated in Figure A.13,

Table A.3 Proposed Numbers and Identities for IMT-2000

Type of Number or Identity	Abbreviation	PLMN Equivalent	ITU-T Standard(s)
IMT-2000 station number	IMISDN	MSISDN	E.164 (revised)/E.213
IMT-2000 user identity	IMUI	IMSI	E.212 (revised)
UPT number	UPTN	None	E.168 (revised)
Personal user identity	PUI	None	E.212 (revised)
Temporary IMT-2000 user Id	TMUI	TMSI	E.213
IMT-2000 station roaming number	IMRN	MSRN	E.213
IMT-2000 global title	IMGT	MGT	E.214/Q.714

and also to support such features as multiple incall UPT registrations on an IMT-2000 mobile station.

In most cases, the existing ITU-T recommendations on numbering and identities will meet the needs of IMT-2000. However, ITU-T is continuously reviewing these recommendations so that they may be enhanced, if necessary, to meet the needs of IMT-2000 as well as other emerging mobility services.

A.7   CONCLUDING REMARKS

This appendix has provided a summary of current standards on numbering and identities and their application to emerging mobile and PCS networks. With the worldwide, explosive demand for mobile and PCS services, support for international roaming and harmonized billing will be a primary requirement, which in turn means that the numbers and identities for these networks must be based on international (ITU-T) standards. The current ITU-T standards on numbering and identities were developed for mobile networks in the 1980s and need to be revised and enhanced to reflect the needs of emerging new systems and services, as well as the new regulatory and competitive environment in which the wireless/PCS services will be offered. The main driving factors for the ongoing activity on numbering and identities in the ITU-T and the various regional and national forums therefore include the following.

Alignment with existing international standards. An example of this type of activity is the current efforts in North America to gradually move from MIN-based identities to E.212-based IMSI in existing AMPS/D-AMPS systems and to use IMSI in emerging new systems like CDMA cellular and PCS.

Support of emerging new systems and services. This type of activity is exemplified by the work in ITU-T SG2 on the revision of international numbering (E.164) and mobile identities (E.212) to accommodate such emerging mobility services as UPT, GMPCS, and IMT-2000. The rapid growth of Internet services and the need for integration and interworking between the numbering and addressing schemes of evolving PSTN/ISDN and PLMN networks and the Internet will pose other challenging tasks.

Accommodate new regulatory and competitive environment. Examples include the ongoing and planned work on *number portability* in North America and Europe and in the ITU-T, which will facilitate movement of mobile/PCS subscribers across service providers and thus promote competition.

A.8   REFERENCES

[01] R. Pandya, "Emerging Mobile and Personal Communication Systems," *Communications Magazine,* Vol. 33, No. 6, June 1995.

[02] R. Pandya, "IMT-2000—Network Aspects," *Personal Communications Magazine,* Vol. 33, No. 7, August 1997.

[03] ITU-T Recommendation E.164 (revised), "The International Public Telecommunication Numbering Plan," International Telecommunication Union, Geneva, May 1997.

[04] ITU-T Recommendation E.212 (revised), "Identification Plan for Mobile Terminals and Mobile Users," International Telecommunication Union, Geneva, November 1998.

[05] ITU-T Recommendation E.213, "Telephone and ISDN Numbering Plan for Land Mobile Stations in Public Land Mobile Networks (PLMN)," *ITU-T Blue Book,* Vol. II, Fascicle II.2, International Telecommunication Union, Geneva, 1988.

[06] ITU-T Recommendation E.214, "Structure of the Land Mobile Global Title for the Signalling Connection Control Part (SCCP)," *ITU-T Blue Book,* Vol. II, Fascicle II.2 International Telecommunication Union, Geneva, 1988.

[07] ITU-T Recommendation E.168 (revised), "Application of E.164 Numbering Plan for UPT," International Telecommunication Union, Geneva, May 1999.

[08] American Telecommunications Industries Solutions, Committee T1 Technical Report #30, "UPT Numbering and Addressing in World Zone 1," ATIS, Washington, DC, 1994.

[09] CIS-3, "TIA Interim Standard on Cellular Mobile Station–Land Station Compatibility," Telecommunications Industries Association, Washington, DC, 1981.

[10] C. I. Cook, "Development of Air Interfaces for PCS," *Personal Communications Magazine,* Vol. 1, No. 4, December 1994.

[11] European Telecommunications Standards Institute, GSM ETR 03.03, "Numbering, Addressing and Identification," ETSI, Sophia Antipolis, France, February 1992.

APPENDIX B

PERFORMANCE BENCHMARKS FOR MOBILE AND PERSONAL COMMUNICATION SYSTEMS AND NETWORKS

B.1 INTRODUCTION

The analog wireless systems were originally targeted for a relatively select group of users, most of whom had mobile telephones installed in their vehicles. The users of the service were willing to trade off the convenience of mobile telephony against service inadequacies in terms of voice quality, call dropoffs, blocking, and delay. However, the demand for mobile and personal communication services has increased dramatically with the availability of low cost, handheld, pocket-sized terminals. These terminals have relatively long battery life and can be used in a variety of environments (in building and pedestrian, as well as in vehicles). With the increasing demand and penetration of wireless services, the emerging innovations in radio technology, and the evolution toward wireless/wireline integration, users of wireless networks now expect quality of service (QOS) and performance comparable to what is available from fixed networks.

The QOS in wireless networks is influenced by a number of factors that are inherent to the use of wireless access and support of terminal mobility. Some of the key factors that influence QOS include:

- available frequency spectrum
- multiplexing and channel assignment methods
- speech encoding and transcoding techniques
- network and signaling architecture for mobility management
- handoff requirements
- authentication and privacy requirements

A closely related factor is system capacity, which also entails potential trade-offs between capacity and performance. Besides the obvious trade-off between traffic capacity (number of users that can be accommodated in a given frequency band for a given cell size) and blocking/delay performance, traffic capacity can be increased by lowering the speech coding rate (e.g., 4 kb/s instead of 8 kb/s coding in a cellular system). Such an increase, however, will come at the expense of lowered speech transmission quality resulting from increased distortion and delay.

Standards that address performance benchmarks for mobile and personal communication networks and network elements are now emerging in such international, regional, and national standards forums as the ITU, ETSI, the ATIS T1 committees, and the TIA. Many of the performance benchmarks for wireless networks and network components are derived from similar benchmarks established for fixed telecommunication networks and associated network components—especially switching systems. These standards need to be reflected in the design of emerging wireless systems and networks so that wireless network operators can deliver improved quality of service to the users in a rapidly growing and competitive market environment.

The focus of this appendix is on descriptions, definitions, and targets or benchmarks for performance parameters relating to the network and network elements. Performance parameters related to radio channel are generally not addressed here. Further, the primary sources of information available in this area and generally used here include relevant ETSI documents on GSM, ITU recommendations, and requirements documents developed by Bellcore. Applicable reference(s) for specific sets of performance benchmarks have been provided. Most of the performance benchmarks given here should be viewed as representative guidelines; individual systems and their implementations may not always conform to these values.

B.2   PERFORMANCE CATEGORIES

In general, telecommunications network and system performance can be categorized using the 3 × 3 matrix shown in Figure B.1 [01], which includes some example performance parameters that are associated with the individual elements of the 3 × 3 matrix. Each row in the matrix represents one of the three basic and distinct communication functions, and each column in the matrix represents one of the three possible outcomes when the communication function is invoked. This matrix model was, however, developed for fixed network applications and primarily addresses functions required for call/connection control. In wireless networks additional (non-call-related)

Performance category / Functions	Speed	Accuracy	Dependability
Access	• Call setup delay • Signaling delay	• Probability of misrouting	• Blocking probability • Connection availability
Information Transfer	• Propagation delay • Data transfer delay	• Distortion • Clipping	• Handoff failure • Probability of call cutoff
Release	• Call release delay	• Charging accuracy	

Figure B.1 A 3 × 3 matrix model for performance categorization.

functions are required for mobility management (e.g., registration, terminal location update, authentication, and encryption). These functions will also influence the system/network performance and the overall QOS perceived by the users.

Whereas the 3 × 3 matrix model is useful in providing a systematic method for identifying and organizing candidate performance parameters, in standards forums performance parameters are generally defined and their target values assigned by partitioning the parameters in the following (more conventional) categories.

Traffic performance. These parameters, which are primarily influenced by the traffic load (under normal operating conditions) and the engineered resources, address the performance during the call setup and release phase. Parameters such as call setup delay, blocking, and handoff failure fall in this category.

Reliability/availability performance. These parameters are influenced by the robustness of the design of hardware and software components in the network. They address long-term performance in terms of both downtime and failure rates, as well as reliability/integrity of system operation in terms of parameters like call cutoff, call misrouting, and incorrect charging.

Transmission performance. These parameters are influenced by the nature of the transmission medium and the nature of multiplexing and encoding techniques deployed. These parameters address performance during the user information transfer phase and include bit error rate, speech channel delay, speech clipping, and quantization distortion performance.

In wireless networks, traffic performance may be influenced by transmission quality. For example, if the interference level (C/I ratio) deteriorates in a given geographic area, there will be increased demand for call handoffs, resulting in a general degradation in call blocking.

B.3   TRAFFIC PERFORMANCE

B.3.1   General Concepts

The need for traffic analysis and engineering arises whenever a system with finite resources is subjected to random service demands. For example, in a wireless network the random service demands are the call originations/ terminations by the subscribers, and the finite resources are the available radio channels, which determine the call blocking (performance). As shown in Figure B.2, the three entities that are central to traffic analysis and design are external traffic load, engineered resources, and observed performance.

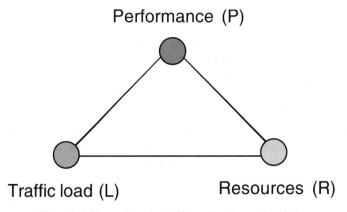

Figure B.2 Key entities for traffic engineering and analysis.

The basic traffic engineering tasks fall into one of the following three categories:

Traffic engineering. Given the performance (benchmark) and the traffic load (Erlangs* CCS/BHCA), determine the amount of required resources (e.g., radio channels, trunks, digit receivers, etc.).

---

* Erlang is a unit of traffic intensity and is defined as the time average for a period T of the number of occupied services

Capacity estimation. Given the amount of available resources and the performance (benchmark), determine the amount of traffic load (Erlangs, CCS, BHCA) that can be carried.

Performance evaluation. Given the amount of resources and the traffic load, evaluate the system performance (delay, blocking).

As a rule, the foregoing tasks are performed by using analytical and/or simulation models to provide a *system* model that includes a load–service curve (throughput curve), which relates performance, traffic load, and resources (Figure B.3).

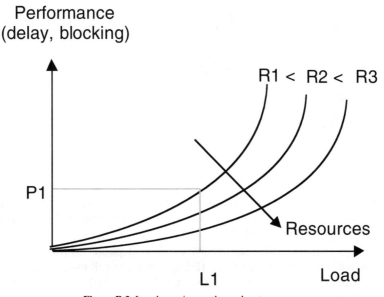

Figure B.3 Load–service or throughput curves.

Since the traffic load represents random service demands, it is necessary to model it in statistical terms. For performance design of network elements (e.g., MSC, BSC, HLR, VLR), reference loads are generally used. As shown in Figure B.4, these reference traffic loads are determined from an event and call model that translates external events (call originations, call terminations, call handoff, etc.) into network traffic loads on control or bearer channels in terms of Erlangs or hundred call seconds (CCS), or processor loads in terms of busy hour call attempts (BHCA).

In wireless networks, the rate of such external events as handoff and location update depends on the radio system design, that is, on the size and configuration of cells and the location area boundaries.

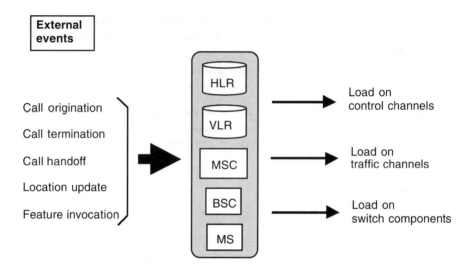

Traffic load (CCS) = attempt rate x holding time (in 100-second units)

Traffic load (Erlangs) = Traffic load (CCS) x 36

Figure B.4 Event and call model for a mobile system.

A single external event like call origination will induce multiple protocol messages on the signaling channel/network and within the individual network elements (BSC, MSC, etc.). The traffic load on the internal components (processors, buffers, etc.) of a network element depends on the call model associated with the network element (NE), where the call model represents the logical flow of events and messages within the NE needed to complete a specific task or subtask.

The standardized reference loads for MSC/BSC and for HLR/VLR are shown in Tables B.1 and B.2, respectively. Though the entries in the two tables are based on the ETSI GSM document [04], the entries for MSC/BSC traffic loads are consistent with ITU-T Recommendation Q.543 [05] for digital ISDN exchanges.

The reference loads in Table B.1 are traffic load conditions under which the performance design objectives are to be met. Reference load A is intended to represent the normal upper mean level of activity that network operators would wish to provide for mobile stations (MS), BS-MSC circuits, and interexchange circuits. Reference load B is intended to represent an increased level beyond normal planned activity levels, which are estimated as 125% of the normal load for Erlangs and 135% of the normal load for BHCA, respec-

Table B.1 Reference Loads for MSC/BSC

Traffic Source	Reference Load A		Reference Load B	
	Erlangs	BHCA	Erlangs	BHCA
Incoming interexchange trunks	0.70	20/trunk	0.85	25/trunk
MS Type W	0.03	1.2	0.0375	1.62
MS Type X	0.06	2.4	0.075	3.24
MS Type Y	0.10	4.0	0.125	5.4

Source: GSM Specification 03.05 [04] and ITU-T Recommendation Q.543 [05].

Table B.2 Reference Loads for HLR/VLR

Transaction Type	Transactions/ Subscriber/Hour	
	HLR	VLR
Call handling	0–4	1–5
Mobility management	1–8	5–8

Source: GSM Specification 03.05 [04].

tively. MS types W, X, and Y represent different application environments that result in different levels of traffic at the mobile station (e.g., light, medium, and heavy users).

If the reference model assumes that significant use is made of supplementary services, the performance of the MSC can be strongly affected, especially when processor capacity can become a limiting item in a given design. The processing delay can be significantly lengthened for a particular call load under such circumstances. The reference model should therefore estimate the fractions of calls that use various supplementary services so that an average processor impact relative to a basic telephone call can be calculated.

The home and visited location registers provide very essential functions in wireless networks in that they maintain permanent and temporary data stores for subscribers roaming within and beyond their home networks. They are key network elements for supporting such mobility management functions as registration, location update, and authentication/ciphering, and in enabling efficient call routing to roaming subscribers.

B.3.2   Traffic Performance Benchmarks

B.3.2.1   Traffic Performance Benchmarks: Network Elements Level

PERFORMANCE BENCHMARKS: MSC/BSC. The traffic performance parameters and targets described here were specified by ETSI for GSM [04]. However, they are based on and are closely aligned with the ITU-T specifications for digital ISDN exchanges [03, 05] for parameter definitions and target values. The following traffic performance parameters have been specified for traffic design of MSC and/or BSC in wireless systems [04]:

- inadequately handled call attempts (blocking)
- user signaling acknowledgment delay
- signaling transfer delay
- through connection delay
- incoming call indication sending delay
- connection release delay
- off-air call-setup (OACSU) delay

Inadequately handled call attempts are attempts that are blocked or are *excessively delayed* within the mobile exchange. *Excessive delays* are defined as delays that are greater than three times the *95th-percentile delay* target.

User signaling acknowledgment delay is the interval from the instant a user signaling message has been received from digital control channel until a message acknowledging the receipt of that message is passed back from the MSC to control channel. Examples of such messages are SETUP ACKNOWLEDGMENT to SETUP, CONNECT ACKNOWLEDGMENT to CONNECT, and RELEASE ACKNOWLEDGMENT to RELEASE.

The MSC signaling transfer delay is the time taken by the MSC to transfer a message from one signaling system to another with minimal or no other exchange actions required. The interval is measured from the instant a message is received from a signaling system until the moment the corresponding message is passed to another signaling system. Examples of messages are ALERT to ADDRESS COMPLETE, ADDRESS COMPLETE to ADDRESS COMPLETE, CONNECT to ANSWER and RELEASE to DISCONNECT.

Through connection delay for originating outgoing traffic is defined as the interval from the instant the signaling information required for setting up a connection through the MSC is received from the incoming signaling system to the instant the transmission path becomes available for carrying traffic between the incoming and outgoing terminations on the MSC.

Through connection delay for internal and terminating traffic is defined as the interval from the instant the CONNECT message is received from the control channel until the through connection is established and available for carrying traffic and the ANSWER and CONNECT ACKNOWLEDGMENT messages have been passed to the appropriate signaling systems.

The incoming call indication sending delay is defined as the interval from the instant at which the necessary signaling information is received from the signaling system to the instant at which the SETUP message is passed to the signaling system of the called subscriber.

Connection release delay is defined as the interval from the instant at which a DISCONNECT or RELEASE message is received from a signaling system until the instant at which the connection is no longer available for use on the call (and is available for use on another call) and a corresponding RELEASE or DISCONNECT message is passed to the other signaling system involved in the connection. The values recommended are shown later (see Table B.9).

Off-air-call-setup (OACSU) delay is the delay in seizing the radio path (traffic channel) after the called subscriber has gone off-hook. It is defined as the interval between receipt of the answer indication from the called subscriber and the instant at which the radio path has been successfully seized.

The target values specified for *inadequately handled call attempts* (blocking) for the MSC are given in Table B.3 and those for the various delay parameters in Table B.4.

Table B.3 Target Values for Inadequately Handled Call Attempts (Blocking) at the MSC

	Inadequately Handled Calls (%)	
Connection Type	Reference Load A	Reference Load B
Internal	1.0	4.0
Originating	0.5	3.0
Terminating	0.2	2.0
Transit	0.19	1.0

Source: GSM Specification 03.05 [04] and ITU-T Recommendation Q.543 [05].

Table B.4 Target Values for Various MSC (Switch) Delay Parameters

Performance Parameter	Target Value (Benchmark)			
	Reference Load A		Reference Load B	
	Mean Value	90th Percentile	Mean Value	90th Pecentile
User signaling acknowledgment delay	200 ms	400 ms	800 ms	1000 ms
Signaling transfer delay	200 ms	400 ms	350 ms	700 ms
Through connection delay (case A)[a]	250 ms	300 ms	400 ms	600 ms
Through connection delay (case B)[b]	350 ms	500 ms	500 ms	600 ms
Incoming call indication delay[c]	600 ms	1000 ms	800 ms	1200 ms
Connection release delay	250 ms	300 ms	400 ms	700 ms
Off-air call setup delay	1000 ms	5000 ms	Not specified	Not specified

[a] Originating outgoing traffic without ancillary function.
[b] Originating outgoing traffic with ancillary function (e.g., echo control).
[c] Assumes en bloc sending in the incoming system.
Source: GSM Specification 03.05 [04] and ITUT Recommendation Q.543 [05].

Performance Benchmarks: HLR/VLR. The relevant delay targets for HLR/VLR are summarized in Table B.5. These performance targets are based on the reference loads for HLR/VLR indicated in Table B.2.

Table B.5 Performance Targets for HLR/VLR

Performance Parameter	Target Values (Benchmarks)	
	HLR	VLR
Delay to retrieve a message	1000 ms	1000 ms
Delay for location registration	2000 ms	2000 ms
Probability of loosing a message	1×10^{-7}	1×10^{-7}

Source: GSM Specification 03.05 [04].

B.3.2.2   *Traffic Performance Benchmarks: Network Level.*   To identify and define pertinent traffic performance parameters at the network level and then assign appropriate targets (benchmarks), it is necessary to understand the detailed call setup and release procedures in the signaling network. Example signaling message flows for a call originating in a mobile network and terminating in the fixed network (mobile orignated call), and for a call originating in the fixed network and terminating in a mobile network (mobile-terminated call) for GSM were shown in Chapter 3 (Figure 3.9 and 3.10, respectively).

The network-level performance is intended to address performance parameters that relate to end-to-end performance, which may directly influence the quality of service perceived by the user. Current mobile network standards primarily focus on traffic and transmission performance. Reliability aspects are covered at the level of individual network elements. The following traffic performance parameters have been specified by the ITU-T for public land mobile networks [06]:

- post selection delay
- answer signal delay
- call release delay
- probability of unsuccessful handoff
- probability of end-to-end blocking

where end-to-end blocking is an aggregate representation of blocking on radio links, blocking on mobile–fixed network links (base station to MSC), and blocking on the fixed (transit) network.

In the majority of wireless calls, a part of the connection will utilize the fixed network over which the mobile system/network designer has little control (e.g., fixed network blocking component in end-to-end blocking). The following definitions and target values for these parameters assume that the mobile and fixed network switches are ISDN capable and that SS7 signaling is deployed in the network. Some of the message types (SETUP, ALERT, CONNECT, RELEASE, etc.) refer to standard ISDN/SS7 signaling messages.

Post selection delay (en bloc sending) is defined as the time interval from the instant of passing the first bit of the initial SETUP message containing all the selection digits by the calling terminal to the access signaling network until the last bit of the first message indicating call disposition has been received by the calling terminal (ALERTING message in the case of a successful call). In the case of mobile-originated calls, the starting instant is the activation of the *send* key by the calling terminal user.

Answer signal delay is defined as the time interval from the instant at which the called terminal passes the first bit of the CONNECT message to its access signaling system until the last bit of the CONNECT message has been received by the calling terminal.

In the PSTN, answer signal delay is the time interval from the called party going off-hook until the ringback tone is removed from the calling terminal. A long answer signal delay can result in the initial part of the speech being clipped at the calling party end.

Call release delay is defined as the time interval from the instant the DIS-CONNECT message is passed, by the user terminal that terminated the call to the access signaling system, until receipt of the RELEASE message by the same terminal (indicating that the terminal can initiate/receive a new call).

Probability of unsuccessful handoff is defined as the probability that a handoff attempt will fail, either because of lack of resources in the target cell or because necessary network resources for establishing the new connection were lacking. The failure condition is based either on a specified time interval since the issuance of the handoff request or on a threshold on signal quality.

Probability of end-to-end blocking is the probability that any mobile-originated/terminated call will be blocked because necessary (radio or land-line) resources are lacking.

The following components of the end-to-end blocking are contributed by the mobile network:

- probability of blocking on the radio channels
- probability of blocking on the mobile-to-fixed network links

The standard target values (benchmark) for the above parameters are summarized in Tables B.6 and B.7 [06].

Table B.6 Target Values: Network-Level Delay Performance

Performance Parameter	Mobile-to-Fixed	Fixed-to-Mobile	Mobile-to-Mobile
Post selection delay	5.5 seconds	9.0 seconds	11.5 seconds
Answer supervision delay	1.0 seconds	1.0 seconds	1.25 seconds
Call release delay	1.0 seconds	1.0 seconds	1.0 seconds

Source: ITU-T Recommendation Q.771 [06].

The values provided in Tables B.6 and B.7 apply under the following assumptions:

- The targets relate to normal traffic loads (e.g., average busy season busy hour load)

Table B.7 Target Values-Network-Level
Blocking Performance

Performance Parameter	All call types (%)
Probability of handoff failure	0.5
Probability of radio channel blocking	1.0
Probability of mobile-to-fixed link blocking	0.5

Source: ITU-T Recommendation Q.771 [06].

- The post selection delay targets include the nominal delays for authentication and ciphering, for paging and alerting, and for routing number transfer. It is also assumed that for mobile-originated calls, the authentication/ciphering procedure is invoked, whereas for mobile-terminated calls it has already been applied.

- The delay targets in Table B.6 are for *local connections* only. Targets applicable to toll and international connections are available in [06].

B.4   RELIABILITY/AVAILABILITY PERFORMANCE

B.4.1   General Concepts and Definitions

Reliability is defined as the probability that a system (or a system element) can perform a required function under stated conditions for a given time interval.

The most important system reliability question involves the uptime of the system. The chance or probability that the system will be up at any particular time, and the percentage of time the system is operating, are two measures that reflect the uptime of the system. If

$$P(t) = \text{probability system is operating at time } t$$

and

$$T(t) = \text{total expected time system has been operating during } [0, t]$$

then two measures of success for any system are given by the limiting values of $P(t)$ or $[T(t)/t]$ (as t becomes very large), which respectively represent the *long-run probability that the system is presently operating* and the *long-run proportion of time the system is operating.*

The two measures actually converge (for large t) and define "*availability*" A for the system. The "*unavailability*" U for the system is simply given by

$$U = 1 - A$$

For hardware reliability, the unavailability metric is commonly transformed to "*downtime*" D, which is defined as the long-run number of minutes per year that the system is unavailable.

Since there are 525,600 minutes in a year, it follows that the downtime D is simply given by

$$D = 525{,}600 \times U$$

"*Failure rate*" F (more correctly, failure intensity) is used to approximate the number of failures expected over any given interval of time.

Whereas downtime parameters are used for reliability requirements that are designed to restrict the unavailability of part or all of a system, the failure rate parameters are used for reliability requirements that are designed to restrict the interruptions received after a stable connection has been established (i.e., to restrict call cutoffs).

Based on the unavailability and failure rate measures defined above, some other useful measures of reliability can be defined/calculated. A commonly used measure of reliability performance is the *mean time to repair* (*MTTR*), which is the long-run average time the system remains in a failed mode before returning to operation. It is given by

$$MTTR = U/F$$

A counterpart to MTTR is *mean time between failures* (*MTBF*), which is the long-run average time the system remains in an operating mode before it fails. The MTBF is given by

$$MTBF = \frac{1 - U}{F}$$

Thus, the reliability performance parameters that are commonly used are:

- availability/unavailability
- down time/failure rate
- MTTR/MTBF

B.4.2 Reliability Performance Benchmarks

B.4.2.1 *Reliability Performance Benchmarks: Network Element Level*

PERFORMANCE BENCHMARKS, MSC/BSC: SYSTEM-LEVEL RELIABILITY The reliability of these entities can be partitioned into "system level" (or hardware) reliability and "call/connection level" reliability. The following reliability performance parameters apply for system-level reliability [07].

Total (system) capability downtime is the expected long-term average annual time spent in failure modes (due to hardware failures and OA&M activities) that affect all lines and trunks. Additionally, loss of signaling capability is considered to cause downtime for all digital trunks. System capability is restored when all lines and trunks have resumed service and the system's engineered call-handling capability has been restored.

Control capability downtime is the expected long-term average annual time spent in failure modes (due to hardware failures and OA&M activities) that affect the control capability of a system (i.e., total loss of local and remote control capability).

Visibility and diagnostic capability downtime is the expected long-term average annual time spent in failure modes (due to hardware failures and OA&M activities) that affect the visibility and/or diagnostic capability of a system (i.e., total loss of local and remote visibility/diagnostic capability).

The performance targets for the (system) reliability parameters just defined are shown in Table B.8. These parameters are limited to hardware reliability that is expected to be built into the system design. Field reliability performance objectives are specified in terms of *downtime performance measurement* (*DPM*) and *outage frequency measurement* (*OFM*) and are generally specified to cover DPM and OFM due to all causes. These are further divided into "supplier attributable" and "service provider attributable." For example, the DPM and OFM for total switch outage are specified as follows:

Overall	**DPM**	**OFM**
	2 min/year	0.1/system/year
Supplier-attributable	1 min/year	0.075/system/year
Service-provider and other	1 min/year	00.025/system/year

Table B.8 System Reliability Performance Targets for MSC/BSC

System Reliability Parameter	Target Value
Total (system) capability downtime	<0.4 min/year
Control capability downtime	<0.4 min/year
Visibility and diagnostics capability downtime	<0.4 min/year
Source: Bellcore GR-512 CORE LSSGR [07].	

PERFORMANCE BENCHMARKS: MSC CALL/CONNECTION LEVEL RELIABILITY.   The performance parameters and targets applicable for call/connection reliability specified for GSM networks are as follows [04]:

Probability of premature call release is the probability that an MSC malfunction will result in the premature release of an established connection in any one-minute interval.

Probability of call release failure is the probability that an MSC malfunction will prevent the required release of a connection.

Probability of misrouting is the probability of a call being misrouted following receipt by the MSC of a valid address.

Probability of no tone is the probability that following receipt of a valid address by the MSC, a call attempt will not encounter tone. (in other words, the caller is left "high and dry").

The recommended target values for these parameters are given in Table B.9.

Table B.9 Performance Targets for Operational Reliability/Integrity for MSC/BSC

Reliability/Integrity Parameter	Target Value (%)
Probability of premature call release	0.002
Probability of call release failure	0.002
Probability of call misrouting	0.01
Probability of no tone	0.002
Source: GSM Specification 03.05 [04].	

PERFORMANCE BENCHMARKS: HLR/VLR RELIABILITY.   The reliability of HLR/VLR may be considered at the same level as the reliability of the service control point in the intelligent networks for database functions. Thus the performance parameters that are applicable to HLR/VLR include platform downtime and probability of lost message.

Platform (hardware) downtime is the inability to support in full or in part the ability of an application process to respond to a call-related message:

Performance target = less than 4.8 hours/year

Probability of a lost message is the probability that a message delivered to HLR/VLR is lost as a result of system malfunction:

Performance target = 1×10^{-7}

Most of the foregoing reliability performance parameters relate to hardware reliability. No standardized software reliability parameter targets are currently available. However, detailed software reliability and quality requirements are used by network operators for testing and acceptance of network equipment [08], [09].

B.4.2.2   Reliability Performance Benchmarks: Network Level. As opposed to traffic performance, where network-level (end-to-end) performance targets have been recommended by standards bodies, no such network-level specifications are currently available for reliability performance of wireless networks. Thus, network-level reliability performance is some aggregate function of the reliability performance of individual network elements.

B.5   TRANSMISSION PERFORMANCE

For public analog and digital (fixed) networks, transmission performance has been specified by the ITU-T in the G-series recommendations [10], [11]. Similar performance targets apply in the case of wireless networks. For wireless networks, however, the existence of multiple radio transmission technologies and multiple access methods has a significant impact on transmission performance.

One of the key transmission performance parameters for emerging digital cellular mobile systems is *speech channel delay*. A significant propagation time between the two ends of a connection causes difficulties in conversation over the connection. This arises from two causes. First, the signal is reflected back from the distant end, causing an echo to the talker. Second, even if ideal echo control were achieved, the delay between a user talking and that party's receipt of a reply from the user at the distant end of the connection could cause conversational difficulty. The delay of the mobile network is made up of the following elements:

- speech transcoding delay
- radio channel coding delay
- mobile network delay (i.e., elements like multiplexing, propagation, switching, echo control)

Round-trip speech channel delay represents the delay introduced by various system components (in the uplink and downlink directions) between the mouth reference point (MRP)/ear reference point (ERP) and the point of interconnection (POI) with the PSTN/ISDN.

Generally a target value for the maximum round-trip speech channel delay is specified along with a delay budget for the major system components.

Table B.10 Recommended Delay Budget for GSM
System Components

TCH/ Full Rate	MSC	BSC	BTS	MS	Total
Downlink	2.0 ms	1.0 ms	70.1 ms	14.3 ms	87.4 ms
Uplink	1.0 ms	1.0 ms	13.3 ms	72.1 ms	87.4 ms

For example, for the GSM system the target for the total delay is 180 ms, and the recommended delay budgets are shown in Table B.10 [04].

Besides the speech channel delay, there are other transmission performance parameters that are relevant for wireless networks, and for which performance targets are available. These include the following:

- echo loss and echo path delay
- quantizing distortion
- loss and loudness rating
- speech clipping

Echo loss and echo path delay indicate the performance of the echo control devices deployed at the point of interconnection (POI) to the ISDN/PSTN (i.e., the MSC). The target values for these parameters, which have been specified by ETSI and ITU-T, are as follows:

$$\text{echo loss} = 45 \text{ dB (minimum)}$$

$$\text{echo path delay} = 60 \text{ ms}$$

Quantizing distortion is introduced when an analog signal is encoded to and from a digital format and is measured in quantization distortion units (QDUs). A QDU is equivalent to the quantization distortion introduced by one average A/D–D/A, 64 kb/s, or A-law PCM codec conforming to ITU-T Recommendation G.711. The following target values apply between the acoustic interface and the POI:

$$\text{quantization distortion for error-free conditions} = 4 \text{ QDU}$$

$$\text{quantization distortion for "realistic" error conditions* } = 7 \text{ QDU}$$

Loss and loudness rating measures the end-to-end loss sustained by an acoustic signal from the mouth of the talker to the ear of the listener, and the

---

* Realistic error conditions are not defind in the standard (G.173) [10].

send loudness rating (SLR) and receive loudness rating (RLR) represent key components of this loss. The target values recommended by ITU-T, which apply up to the POI, are as follows:

$$\text{SLR} = +7 \text{ dB to } +9 \text{ dB}$$

$$\text{RLR} = +1 \text{ dB to } +3 \text{ dB}$$

Note that ratings for *hands-free* sets will be higher.

Speech clipping represents the loss of the start or end of a speech burst. The main cause of speech clipping is the use of voice switching controlled by voice activity detection (e.g., by echo cancelers located at the MSC). Clipping in this context does not refer to lost frames due to burst errors in the radio channel.

Voice switching may occur in devices within the network or within terminal devices. The following devices employ voice switching:

- echo suppressers
- echo cancelers (with center clippers)
- digital speech interpolators (DSI)
- discontinuous transmission (DTx)
- loudspeaking (hands-free) telephones

The target values for speech clipping performance recommended by the ITU-T [10], are as follows:

Duration of clipping	<64 ms
Frequency of speech clipping	<0.002 × speech activity factor
Percentage of speech clipping	<0.2%

B.6   CONCLUDING REMARKS

Until recently, mobile network users were willing to trade off quality of service and performance against the convenience of terminal mobility and the resulting ability to originate and receive calls while on the move. However, as the demand and penetration of wireless services increases at a very rapid rate, and efforts to integrate wireless and wireline networks and services becomes reality, transparency of user-perceived performance between wireless and wireline access will become critical. Thus it is imperative that wireless networks and systems be designed and dimensioned to standard benchmarks that are as close as possible to those used in the fixed networks.

This appendix has attempted to summarize the international and regional standards that are currently available for dimensioning wireless networks and for the design of wireless network elements. Development of standards for wireless networks (as in other areas) is an ongoing process. The standards are continuously being reviewed, revised, and enhanced to keep pace with the wireless market/industry demands and with evolving radio and network technologies.

B.7 REFERENCES

[01] R. Pandya, K. Basu, and S. Tseng, "Some Performance Benchmarks for the Design of Wireless Systems and Networks," *Proceedings of the 15th International Teletraffic Congress—ITC 15,* (Washington, DC, June 22–27, 1997), Elsevier, New York, 1997, pp. 243–254.

[02] O. Avellaneda and R. Pandya, "Traffic Grade of Service Standards for Cellular Mobile Radio Systems," International Teletraffic Congress, Turin, Italy, June 1988.

[03] R. Pandya et al., "Traffic Modeling of a Cellular Mobile System," International Teletraffic Congress, Kyoto, Japan, September 1985.

[04] European Telecommunications Standards Institute, GSM Specification 03.05, "Technical Performance Requirements," ETSI, Sophia Antipolis, France, October 1993.

[05] ITU-T Recommendation Q.543, "Digital Exchange Performance Design Objectives," *ITU-T Blue Book,* Vol. VI, Fascile VI.5, International Telecommunication Union, Geneva, 1984.

[06] ITU-T Recommendation E.771, "Network Grade of Service Parameters and Target Values for Circuit Switched Public Land Mobile Services," International Telecommunication Union, Geneva, 1996.

[07] Bellcore GR-512-CORE, LSSGR, Issue 1, "Local Access Transport Area (LATA) Switching System Generic Requirements for Reliability," Bellcore, Navasink, NJ, January 1995.

[08] Bellcore TR-NWT-000179, Issue 2, "Quality System Generic Requirements for Software," Bellcore, Navasink, NJ, June 1993.

[09] GR-282-CORE, "Software Reliability and Quality Acceptance Criteria," Bellcore, Navasink, NJ, December 1994.

[10] ITU-T Recommendation G.173, "Transmission Planning Aspects of the Speech Service in Digital Public Land Mobile Networks," International Telecommunication Union, Geneva, March 1993.

[11] ITU-T Recommendation G.174, Transmission Performance Objectives for Terrestrial Digital Wireless Systems Using Portable Terminals to Access the PSTN," International Telecommunication Union, Geneva, June 1994.

[12] European Telecommunications Standards Institute, GSM Specification 03.50, "Transmission Planning Aspects of Speech Service in the GSM PLMN System, ETSI," Sophia Antipolis, France, August 1995.

APPENDIX **C**

LIST OF
ABBREVIATIONS AND
ACRONYMS USED IN
THE BOOK

3GPP	Third-Generation Partnership Project
AAL	ATM adaptation layer
ABR	available bit rate
ABSBH	average busy season busy hour
AC	alerting channel
AC	authentication center
ACH	access response channel
ACTS	Advanced Communication Technologies and Services (Europe)
ACW	address code word
A/D	analog to digital
ADPCM	adaptive differential pulse code modulation
ADS	asynchronous data service
AGC	automatic gain control
AGCH	access grant channel
AKA	authentication and key agreement (PACS)
ALT	automatic link transfer (PACS)
AMF	authentication management function
AMP	acknowledgment Mode Protocol (PACS)
AMPS	advanced mobile phone system
AMSC	anchor mobile switching center
AMTS	automated mobile telephone system

ANSI	American National Standards Institute
ANSI-41	American National Standards Institute-41 (aka TIA IS-41)
AP	access point
ARA	alerting/registration channel (PACS)
ARC	alerting request channel
ARF	access link relay function
ARIB	Association of Radio Industries and Businesses (Japan)
ARQ	automatic repeat request
ATDMA	asynchronous time division multiple access
ATIS	Alliance for Telecommunication Industry Standards (United States)
ATM	asynchronous transfer mode

BCCH	broadcast control channel
BCH	Bose–Chaudhary (encoding)
BER	bit error rate
BHCA	busy hour call attempts
B-ISDN	broadband integrated digital services network
BPSK	binary phase shift keying
BRI	basic rate interface
BS	base station
BSC	base station controller
BSMAP	base station mobile application part
BSS	basic service set (IEEE 802.11 WLAN Specification)
BSS	base station system
BSSAP	BSS Application Protocol
BSSGP	BSS GPRS Protocol
BTA	basic trading area
BTS	base transceiver station
BTSM	BTS management
BW	bandwidth

CA	collision avoidance
CAC	common access channel (PDC)

CAC	channel access control
CAI	common air interface
CAMEL	customized application for mobile network enhanced logic
CATT	China Academy of Telecommunications Technology
CAVE	cellular authentication and voice encryption
CBR	constant bit rate
CC	call control
CC	country code
CCC	cordless cluster controller (DECT)
CCAF	call control agent function
CCBS	call completion to busy subscribers
CCCH	common control channel
CCF	call control function
CCFP	common control fixed part (DECT)
CCH	control channel
CCS	common channel signaling
CCS	hundred call seconds
CD	collision detection
CDCS	continuous dynamic channel selection (DECT)
CDMA	code division multiple access
CDPD	cellular digital packet data
CEPT	Conference of European Postal and Telecommunications Administrations
CHM	channel marker
CHMF	channel marker—fixed
CHMP	channel marker—portable
C/I (ratio)	carrier-to-interference ratio
CI	continuity indicator
CLNP	Connectionless Network Protocol (OSI)
CLNS	Connectionless Network Service
CLPC	closed-loop power control
CM	connection management
CN	core network
CnCAF	connection control agent function
CnCF	connection control function
CO	central office
Codec	coder–decoder

CRC	cyclic redundancy check
CS	capability set
CS	cell station (PHS)
CSAP	channel access control sublayer, service access point
CSMA/CA	carrier sense multiple access with collision avoidance
CS-*n*	Capability Set *n* (IMT-2000)
CSP	channel status protocol (CDPD)
CSPDN	circuit-switched public data network
CT2	Cordless Telephony 2
CTA	cordless terminal adapter
CTIA	Cellular Telecommunication Industry Association
CTM	cordless telephone mobility
CTS	clear to send
CUG	closed user group
CUSF	call-unrelated service function
CW	contention window

D-AMPS	digital advanced mobile phone system
DCC	digital control channel (D-AMPS)
DCCH	dedicated control channel (GSM)
DCF	distributed coordination function (IEEE 802.11)
DCS1800	digital cellular system, 1800 MHz
DCW	data code word
DECT	Digital Enhanced Cordless Telecommunications (aka Digital European Cordless Telecommunications)
DHCP	dynamic host configuration protocol
DIFS	distributed coordination (DCF) function initial interframe space (IFS)
DLC	data link control
DMSC	drift mobile switching center
DN	directory number
DPM	downtime performance measurement
DQPSK	differential encoded quadrature phase shift keying
DPRS	DECT packet radio service
DS	direct sequence

DS	distribution system (IEEE 802.11)
DSI	digital speech interpolator
DSL	digital subscriber loop
DSMA	data sense multiple access
DSP	digital signal processing
DSSS	direct sequence spread spectrum
DTAP	direct transfer application protocol
DTMF	dual-tone multifrequency
DTx	discontinuous transmission

E-BCCH	extended broadcast channel
EDGE	enhanced data rates for GSM evolution
EFR	enhanced full rate (coder)
EIA	Electronic Industries Association
EIR	equipment identity register
EO	end office
EOC	embedded operations channel
ERP	ear reference point
ERP	effective radiated power
ES	earth station
ESN	electronic serial number
ETSI	European Telecommunications Standards Institute
EVRC	enhanced variable-rate codec

FA	foreign agent (mobile IP)
FAC	final assembly code
FACCH	fast associated control channel
F-BCCH	fast broadcast channel
Fc	carrier frequency
FCC	Federal Communications Commission (U.S.A.)
FCCH	frequency correction channel
FDD	frequency division duplex
FDMA	frequency division multiple access
FE	functional entity

FEA	functional entity action
FEC	forward error correction
F-ES	fixed-end system (CDPD)
FFSK	fast frequency shift keying
FH	frequency hopping
FHSS	frequency-hopping spread spectrum
FM	frequency modulation
FPLMTS	Future Public Land Mobile Telecommunication Systems (now IMT-2000)
FR	frame relay
FrSw	frame switch
FSK	frequency shift keying
FT	format type
FWA	fixed wireless access

GAP	generic access profile (DECT)
GFSK	Gaussian frequency shift keying
GEO	geostationary (satellites)
GGSN	gateway GPRS support node
GIP	GSM interface profile (DECT)
GLR	gateway location register (PDC system)
GMCC	gateway mobile control center (PDC)
GMPCS	global mobile personal communication by satellite
GMSC	gateway mobile switching center
GMSK	Gaussian minimum shift keying
GMSS	global mobile satellite system
GOCC	ground operation control center
GPCF	geographic position control function
GPF	geographic position function
GPRS	general radio packet service (GSM)
GSA	geographic service area
GSM	Global System for Mobile communication (previously Groupe Special Mobile)
GSN	GPRS support node
GTP	GPRS tunnel protocol

GTT	global title translation
GW	gateway

HA	home agent (mobile IP)
HDLC	high level data link control
HDR	high data rate
HDSL	high speed digital subscriber loop
H-GMSC	home—gateway MSC
HGSN	home GPRS support node
HIPERLAN	High Performance European Radio LAN
HLR	home location register
HSCSD	high speed circuit-switched data

IAF	intelligent access function
IC	integrated circuit
ICMP	Internet Control Message Protocol
ICO	Inclined Circular Orbit (satellites)
IE	information element
IEEE	Institute of Electrical and Electronics Engineers
IETF	Internet Engineering Task Force
i/f	interface
IFS	initial interframe space
IGS	interworking gateway switch (PDC system)
IMEI	international mobile equipment identity
IMGT	IMT-2000 global title
IMRN	IMT-2000 routing number
IMSI	international mobile station identity
IMT-2000	International Mobile Telecommunications 2000 (formerly FPLMTS)
IMUI	IMT-2000 user identity
IN	intelligent network
INC	industry numbering committee
IP	intelligent peripheral
IP	Internet Protocol
IS	Interim Standard (TIA/ANSI)

IS	intermediate system (CDPD)
ISC	international switching center
ISDN	integrated services digital network
ISL	inter satellite link
ISM	industrial, scientific, and medical (frequency band in the United States)
ISU	Iridium subscriber unit
ITU	International Telecommunication Union (UN)
ITU-R	ITU radio communications sector
ITU-T	ITU telecommunications standardization sector
IU	interface unit
IUT	ICO user terminal
IW	interworking (unit)
IWF	interworking function

Kc	ciphering key
Ki	secret key
Ksps	kilo symbols per second

LAES	lawfully authorized electronic surveillance
LAI	location area identity
LAN	local-area network
LAPC	link access protocol C
LAPD	link access protocol for D channel
LAPR	link access protocol for radio (CT2)
Lb	link protocol (DECT)
LBR	low bit rate
LEC	local exchange carrier (United States)
LEO	low earth orbit (satellites)
LID	link identity
LLC	logical link control
LMF	location management function
LR	location register

LATA	local access transport area
LSSGR	LATA switching system generic requirements (Bellcore)

MAC	media access control
MAC	message authentication code
MAP	Mobile Application Protocol
MCC	mobile countrol center (PDC system)
MCC	mobile country code
MCF	mobile control function
Mcps	mega chips per second
MDBS	mobile data base station (CDPD)
MDLP	mobile data link protocol (CDPD)
MD-IS	mobile data—intermediate system (CDPD)
MEO	medium earth orbit (satellites)
M-ES	mobile-end system (CDPD)
MFA	manufacturer (code)
MGE	management entity
MGPF	mobile geographic position function
MGT	mobile global title
MHF	mobile home function (CDPD)
MIN	mobile identification number
MM	mobility management
MMAP	multimedia access profile (DECT)
MN	mobile node (mobile IP)
MNC	mobile network code
MNP	mobile network protocol
MOU	memorandum of understanding
MRP	mouth reference point
MRTR	mobile radio transmission and reception
MS	mobile station
MSC	mobile switching center
MSF	mobile serving function (CDPD)
MSIN	mobile station identification number
MSISDN	mobile station ISDN number
MSRN	mobile station routing number

MSS	mobile satellite service
MT	mobile terminal
MTA	major trading area
MTBF	mean time between failures
MTP	message transfer part (SS7)
MTTR	mean time to repair

N.A.	North America
NAMPS	narrowband AMPS
NANC	North American Numbering Council
NANP	North American Numbering Plan
NANPA	North American Numbering Plan Administrator
NAP	network access point
NAV	net allocation vector
NC	network code
NCC	network control center
NDC	network destination code
NE	network element
NID	network identification number
NMSI	national mobile station identity
NMT	Nordic Mobile Telephone
NNI	network to network interface or node-to-node interface
NOP	network and operations plan (PCS)
NPA	numbering plan area
NSN	national significant number
NT	network termination
NT	nontransparent
NTT	Nippon Telephone and Telegraph

OA&M	operations, administration, and maintenance
OACSU	off-air call setup
OFDM	orthogonal frequency division multiplex
OFM	outage frequency measurement
OLPC	open-loop power control

OQPSK	offset quadrature phase shift keying
OS	operations system
OSI	open system interconnection
OTA	over-the-air activation
OTASP	over-the-air service provisioning

PACA	priority access and channel assignment
PACS	personal access communication system
PAD	packet assembly and disassembly
PAMR	public access mobile radio
PAP	public access profile (DECT)
PC	point code (SS7)
PCF	point coordination function (IEEE 802.11)
PCH	paging channel
PCI	personal communications interface
PCIA	Personal Communications Industry Association
PCM	pulse code modulation
PCMCIA	Personal Computer Memory Card International Association
PCS	personal communication service
PCS1900	personal communication system, 1900 MHz
PDA	personal digital assistant
PDC	Personal Digital Cellular (Japan)
PDSN	packet data support node
PDTCH	packet data traffic channel
PDU	protocol data unit
PE	physical entity
PGW	packet gateway module (PDC)
PHL	physical layer
PHS	Personal Handyphone System (Japan)
PID	portable identity
PIFS	point coordination function (PCF) initial interframe space (IFS)
PIN	personal identification number
PLMN	public land mobile network
PMR	private mobile radio
PN (code)	pseudo–random noise (code)

PNNI	private network-to-network interface (ATM)
POI	point of interconnection
POS	point of sale (terminals)
PPM	packet processing module (PDC)
PPP	point-to-point protocol
PRC	priority request channel (PACS)
PRMA	packet reservation multiple access
PS	PHS station
PSCAF	packet service control agent function
PSCF	packet service control function
PSGCF	packet service gateway control function
PSK	phase shift keying
PSPDN	packet-switched public data network
PSTN	public switched telephone network
PTM	point-to-multipoint
PTP	point-to-point
PUI	personal user identity
PVC	private virtual circuit

QAM	quadrature amplitude modulation
QCELP	quadrature code excited linear predictive (coder)
QDU	quantization distortion unit
QI	quality indicator
QPSK	quadrature phase shift keying
QOS	quality of service
QSAFA	quasi-static automatic frequency assignment (PACS)

RA	rate adaption
RA	registration area
RACF	random access control function
RACH	random access control channel
RAL	radio access layer (WATM)
RAN	radio access network
RAND	random number

RCC	radio control channel
RCC	radio common carrier
RCF	radio control function
RCR	Research and Development Center for Radio Systems (Japan)
RF	radio frequency
RFP	radio fixed part (DECT)
RFTR	radio frequency transmission and reception
RLC	radio link control
RLP	Radio Link Protocol
RLR	receive loudness rating
RNC	radio network controller
RP	radio port
RPCU	radio port control unit
RPE/LTE	residual pulse excitation/long-term prediction (vocoder)
RR	radio resource (management)
RRSI	received signal strength indicator
R-S	Reed–Solomon (encoding)
RTF	radio terminal function
RTS	request to send
RTT	radio transmission technology

SAAL	signaling for AAL (ATM adaptation layer)
SAC	service access code
SACF	service access control function
SACHH	slow associated control channel
SAN	satellite access node
SAP	service access point
SAT	supervisory audio tone
SBC	system broadcast channel
S-BCCH	SMS broadcast channel
SC	service center
SCC	satellite control center
SCCH	synchronization channel
SCCP	signaling connection control part (SS7)
SCEF	service creation environment function

SCF	service control function
SCH	shared channel
SCP	service control point
SCUAF	service control user agent function
SDCCH	stand-alone dedicated control channel
SDF	service data function
SDL	system description language
SDMA	space division multiple access
SDO	standards development organization
SDP	service data point
SDU	signaling data unit
SG	Study Group (ITU)
SGSN	serving GPRS support node
SIBF	system access information broadcast function
SIC	system information channel (PACS)
SIB	service-independent building block
SID	system identification number
SIFS	short initial interframe space (IFS)
SIM	subscriber identity module (GSM)
SINR	signal-to-interference/noise ratio
SLP	service logic program
SLR	send loudness rating
SMAF	service management agent function
SMF	service management function
SMG	Special Mobile Group (ETSI)
SMR	specialized mobile radio
SMS	short message service
SMS-SC	SMS-service center
SMS-GW	SMS-Gateway
SN	service node
SN	subscriber number
SNCF	satellite network control function
SNDCP	sub network-dependent convergence protocol (CDPD)
SNR	signal-to-noise ratio
SOCC	satellite operation control center
SP	service profile

SP	service provider
SP	Standards Project (TIA)
SPAN	Services and Protocols for Advanced Networks (ETSI Technical Committee)
SPI	service provider indicator
SREJ	selective reject
SRES	signed response
SRF	specialized resource function
SRTT	sets of RTTs
SRU	speech recognition unit
SS	supplementary services
SS7	Signaling System 7
SSD	shared secret data
SSF	service switching function
SSN	subsystem number (SS7)
SSP	service switching point
STP	signaling transfer point (SS7)
SU	subscriber unit
SV	space vehicle (satellite)

TAC	type approval code
TACS	total access communication system
TCAP	Transaction Capability Application Protocol (SS7)
TCH	traffic channel
TCP	transmission control protocol
TDD	time division duplex
TDM	time division multiplex
TDMA	time division multiple access
TEI	terminal equipment identity
TETRA	terrestrial trunked radio
TG8/1	Task Group 1 of ITU-R Study Group 8 (responsible for IMT-2000)
TIA	Telecommunications Industry Association (U.S.A.)
TID	terminal identity

TLDN	temporary local directory number
TMSI	temporary mobile station identity
TMUI	temporary IMT-2000 user identity
TP	transfer protocol
TRAU	transcoder and rate adaption unit
TSB	Telecommunications Standardization Bureau (ITU)
TST	time slot transfer
TTA	Telecommunications Technology Association (South Korea)
TT&C	telemetry, tracking, and control
TTC	Telecommunications Technical Committee (Japan)
TUP	telephone user part (SS7)

UBR	universal bit rate
UDP	User Datagram Protocol
UIM	user identity module
UMTS	Universal Mobile Telecommunication System (ETSI)
UNI	user–network interface
U-NNI	unlicensed national information infrastructure (spectrum for)
UPC	user-specific control channel (PDC)
UPCH	user packet channel
UPR	user performance requirements
UPT	universal personal telecommunication
UPTAC	UPT access code
UPTAN	UPT access number
UPTGT	UPT global title
UPTI	UPT indicator
UPTN	UPT number
UTRA	UMTS terrestrial radio access
UWCC	Universal Wireless Communications Consortium

VAD	voice activity detection
VBR	variable bit rate
VC	virtual circuit

VCI   virtual circuit identity
VHE   virtual home environment
VLR   visitor location register
VMCC   visited mobile control center (PDC)
VMSC   visited mobile switching center
VSELP   voice sum excited linear predictive (vocoder)

WATM   wireless ATM (asynchronous transfer mode)
WACS   wireless access communication system
WARC   World Administrative Radio Council
WB-DS   wideband—direct sequence (CDMA)
WCC   wireline common carrier
WCPE   wireless customer premises equipment
WEI   word error indicator
WEP   wired equivalency privacy
WIMS   wireless multimedia and messaging service
WIN   wireless IN (intelligent network)
WLAN   wireless local-area network
WRC   World Radio Conference
WWW   World Wide Web
WZ1   World Zone 1 (North American countries allocated country
                 code 1)

INDEX

Page numbers followed by "f" indicate figures; page numbers followed by "t" indicate tables.

ABOUT THE AUTHOR

Raj Pandya is a consultant on mobile communications with a primary focus on network standards for IMT-2000—the international standard for the next-generation mobile communication system being specified by the International Telecommunications Union (ITU). He is the chairman of Working Party 3 in ITU-T Study Group 11 responsible for developing signaling requirements for IMT-2000.

Dr. Pandya has worked in the Systems Engineering Division of Bell-Northern Research and Nortel Networks for over 20 years. During this period, his research activities included teletraffic modeling of telecommunication systems and networks, mobile network modeling and performance, and personal communication services (PCS). Since 1980 he has been active in the telecommunications standards activities of the ITU, where he has chaired working groups on areas such as: international numbering plans; ISDN traffic performance; network capabilities to support UPT; signaling and switching requirements for current and future mobile systems; and traffic engineering and performance for mobile networks. From 1990 to 1992, he was on assignment as acting director of the Teletraffic Research Center, University of Adelaide, Australia.

Dr. Pandya has published numerous papers on performance modeling of communications networks and emerging wireless network technologies and standards. He has taught undergraduate and graduate courses on communications engineering in India and Canada. Dr. Panyda holds a Master's degree in radio physics and electronics from the University of Calcutta, India; a Master's degree in electrical engineering from the University of Toronto, Canada; and a Ph.D. in electrical engineering from Carleton University, Ottawa, Canada.